Plastics Institute of America, Inc.
A Non-Profit, Educational, and Research Organization

FOODPLAS VIII-91

Plastics in Food Packaging

Proceedings from The Eighth Annual FoodPlas Conference

March 5–7, 1991
Orlando, Florida

FOOOPLAS VIII-91
a**TECHNOMIC**⁵publication

Published in the Western Hemisphere by
Technomic Publishing Company, Inc.
851 New Holland Avenue
Box 3535
Lancaster, Pennsylvania 17604 U.S.A.

Distributed in the Rest of the World by
Technomic Publishing AG

Printed in the United States of America
10 9 8 7 6 5 4 3 2 1

Main entry under title:
    FOODPLAS VIII-91: Plastics in Food Packaging

A Technomic Publishing Company book
Bibliography: p.

ISSN No. 1048-2016
ISBN No. 87762-866-1

TABLE OF CONTENTS

SESSION 1

HOW SUPERMARKETS LOOK AT PLASTICS

PLASTICS AND THE PURCHASING EQUATION

Mona Doyle, President
The Consumer Network
3624 Market Street
Philadelphia, PA 19104

Biography

Mona Doyle is president of The Consumer Network, Inc., a consumer research and marketing firm that specializes in the food industry.

The Consumer Network operates a national panel of vocal supermarket shoppers who represent a leading edge of the food buying marketplace. The firm's clients include major producers as well as advertising agencies, packagers, and retailers. Consumer Network services include marketing communications counsel; focus group and other qualitative research; market analysis; and customer focusing.

Before founding The Consumer Network, Mona served as Vice President, Consumer Affairs and Research for the Pantry Pride/Food Fair supermarket chain. She previously held research and marketing positions with RCA, The Bell Telephone Company of Pennsylvania, and the Arthur S. Kranzley consulting organization which developed computer software for banks and retailers.

Mona appears on national television shows including Face the Nation and writes monthly columns for several trade publications. She is frequently quoted in Advertising Age,

Business Week, Changing Times, and major newspapers including the New York Times, USA Today, the Wall Street Journal, the Los Angeles Times and the Chicago Tribune.

A psychology honors graduate of the University of Pennsylvania, she has done graduate work in business at Temple and Drexel Universities and completed the Cornell Food Executive Program.

Her current board memberships include the Better Business Bureau of Eastern Pennsylvania; the New Jersey Board of Pharmacy; and the Executive Committee of the Multiple Sclerosis Society. She is a board certified member of the Institute of Management Consultants (CMC); a professional member of the Qualitative Research Society of America; and a founding member of Philadelphia's Forum of Executive Women and the National Society of Consumer Affairs Professionals in Business.

Abstract

Supermarkets are caught in a Plastic bind.

They see benefits in plastic that are meaningful to their performance and their operating costs.

They see defense of plastic to their consumers as frequently self-defeating. Even supermarkets with super-high credibility ratings get in consumer hot water from defending plastics.

The Facts aren't The Truth

There are multiple sets of pertinent facts.

Facts that mean something now (next week, next month, next year) to customers.

Facts that mean something now (next week, next month, next year) to employees.

Facts that mean something to management.

Perception has power

But they keep coming back to the fact that plastic has performance advantages.

Efficiency, reliability, profit, space, etc.

EXECUTIVE/BUYING MANAGEMENT
QUALITATIVE ASSESSMENT: ATTRIBUTES AND ATTITUDES

+   FOR PRODUCT PRESENTATION

+   FOR STORAGE

+   FOR THROUGHPUT

+   FOR PRODUCT PROTECTION

+   FOR SPACE

+   FOR SHRINK

=   FOR LABOR

STORE MANAGEMENT
NUMERIC RATINGS

A*	B*	
4.0	28%	FOR SHIPPING
3.6	11%	FOR STORAGE
3.6	28%	FOR SAFETY - BIG VARIATION HERE
3.6	17%	FOR SHELF LIFE
3.5	22%	FOR VALUE
3.1	0	FOR TURNOVER
2.7	6%	FOR RECYCLING
2.5	11%	FOR SOLID WASTE

A* = Mean rating on 5 point scale

B* = Percent rating 5, excellent

CHANGING VIEWS OF PLASTIC

Supermarket managers and executives see plastics as environmental negatives <u>and</u> as materials of choice for packaging.

From the retailers' side, plastic packaging appears to deliver convenience and value to consumers and advantages in distribution, storage, safety, and shelf life for the retailer. Giving up plastic would be taking steps backward. No wonder many retailers are actively supporting plastic recycling efforts!

We asked consumers to obtain and comment on supermarket managers' opinions about plastic. Most of the consumumers who were able to get their managers to respond said they agreed with the managers' evaluations up and down the line. But a significant number expressed surprise at how highly the managers rated plastic. And one said the survey made her realize that supermarkets had been fighting bottle bills for years but were supporting efforts to take back plastic.

The role of plastic in the purchase decision has shifted and diminished in recent months. More consumers are learning that plastic -- even styrofoam -- may not be as bad as they thought. And more are reacting to perceived excesses in packaging -- paper and plastic both perceived as environmental problem posers -- and less packaging and more recycling perceived as solutions.

Between 5% and 7% of our consumers are reacting to obiously plastic packaging as a barrier to purchase. That barrier operates as a red flag -- a reason to reconsider and see if there is an environmentally better alternative that is as convenienct and no more expensive. If no "good" alternatives are found, the item in plastic is purchased. It's not the full stop that results in giving up soup because it's salty or giving up eggs or cheese because of cholesterol.

A total of 15% to 18% think about plastic as top of mind issue in packaging, but not all of this thinking is environmental. Some continue to associate plastic with convenience; others with the opening or reclosing difficulties; and others with double packaging or layered packaging. Double layers of packaging appear more enviornmentally unfriendly than single layers -- alternative layers of paperboard and plastic look worse than plastic alone!

PLASTICS
THE GOOD, THE BAD AND THE OPPORTUNITY
Jon Seltzer
Director of Corporate Planning
Super Valu Stores, Inc.
P.O. Box 990
Minneapolis, MN 55400

Biography

Jon Seltzer is Director of Corporate Planning at Super Valu Stores, Inc. in Minneapolis, Minnesota, where for the past 10 years he has been involved with the long term direction of the company and its government relations and community activities. Prior to joining Super Valu, Jon worked as a business planner at Clark Equipment after receiving his masters of business administration from the University of Chicago. After graduating from Lawrence University in Appleton, Wisconsin, Jon spent four years in the Peace Corps working in Tunisia and the Solomon Islands.

In addition, Jon is on the board of the Minnesota Charities Review Council; serves on the executive committee of the Minnesota Minority Education Partnership; is a Food Industry Crusade Against Hunger trustee; and is currently serving as chair of the Food Marketing Institute government relations committee.

Abstract

In an uncertain world, one thing is certain -- a discussion of plastics and the environment will be a spirited one.

Plastics are the good, the bad and an opportunity. Plastics provide many benefits as a packaging material. These same attributes are the bad and represent, at least in the consuming public's eye, the downside to plastic. It is my opinion, plastics will continue to play a major role in packaging and this represents the opportunity for those plastic manufacturers and suppliers who can address the environmental challenges of the 90s.

Three separate trade offs, or competing interests, must be balanced in a discussion of plastics and the environment.

- First of all, the trade off between protecting food products and the cost of providing this protection relative to the environmental trade off.

- Second, consumers - desires for convenience, timesaving versus the environmental trade off.

- And three, retailers desires to provide service departments and the environmental trade off.

A combination of these three sets of trade offs both provide strong support for plastics continued role as a packaging material in the years to come and underscore the importance of addressing environmental concerns.

The EPA's waste hierarchy reduce, reuse, recycle provide insight as to the path the food industry must take to address this issue. The key question of how to pay for this is where agreement breaks down.

It is my contention, and that of many other retailers, wholesalers and consumers, that the plastic industry, the producers of plastic, need to do more to provide a market for recycled plastic products. Suppliers who seek to resolve consumers, retailers and package food manufacturers plastic packaging recycling concerns will have a competitive advantage in the years ahead.

PLASTICS, THE GOOD, BAD AND THE OPPORTUNITY

BY

JOHN SELTZER, DIRECTOR

CORPORATE PLANNING

SUPER VALU

SLIDE I	PLASTICS - THE GOOD, THE BAD & THE OPPORTUNITY
SLIDE 2	SUPER VALUE - THE RETAIL SUPPORT CO.
SLIDE 3	SUPER VALUE - RETAIL SUPPORT OPERATING TERRITORY
SLIDE 4	PERSPECTIVE OF A RETAILER
SLIDE 5	PERCEPTION IS REALITY
SLIDE 6	TWO RULES IN RETAILING
SLIDE 7	TWO RULES IN RETAILING 1. THE CUSTOMER IS ALWAYS RIGHT
SLIDE 8	TWO RULES IN RETAILING 1. THE CUSTOMER IS ALWAYS RIGHT 2. SEE RULE NUMBER ONE
SLIDE 9	CUSTOMERS ARE VERY INTELLIGENT AND CAPABLE BUYERS. ULTIMATELY THEY WILL BUY FROM THE SOURCE THAT DOES THE BEST JOB FOR THEM. THAT'S THE WAY FREE ENTERPRISE SYSTEM WORKS.
SLIDE 10	GIVEN THE TWO RULES OF RETAILING...CHANGE THE CUSTOMERS' PERCEPTIONS
SLIDE 11	GIVEN THE TWO RULES OF RETAILING...CHANGE THE CUSTOMER'S PERCEPTIONS. PLASTIC IS NOT THE PROBLEM

SLIDE 12	PLASTIC IS NOT THE OVERWHELMING PRESENCE IT'S CRACKED UP TO BE. ALL PLASTICS ACCOUNT FOR ONLY 7 PERCENT OF MUNICIPAL SOLID WASTES, AND POLYSTYRENE - PUBLIC ENEMY NUMBER ONE IN THE ENVIRONMENTALIST'S VIEW - ONLY A FEW TENTHS OF ONE PERCENT. FURTHERMORE, THE FACT THAT PLASTIC DOESN'T BIODEGRADE IS AS MUCH A VIRTUE AS A SIN. IT MEANS THAT PLASTIC DOESN'T RELEASE CHEMICALS INTO THE GROUNDWATER
SLIDE 13	GIVEN THE TWO RULES OF RETAILING...CHANGE THE CUSTOMERS' PERCEPTIONS. PLASTICS IS RECYCLABLE
SLIDE 14	THE PLASTICS INDUSTRY HAS RESPONDED WITH PR STUNTS AND PARK BENCHES THE STARS OF THE "FEEL GOOD ABOUT PLASTICS" BROCHURES. THIS HAS LED TO WIDESPREAD SPECULATION AMONG PRESTIGIOUS PHYSICISTS AS TO HOW MANY PLASTIC PARK BENCHES THE EARTH'S CRUST CAN SUPPORT. EIGHT MILLION TONS OF PLASTIC PACKAGING PER YEAR TRANSLATES INTO 128 MILLION FOUR-FOOT PARK BENCHES EACH YEAR.
SLIDE 15	GIVEN THE TWO RULES OF RETAILING
SLIDE 16	GIVEN THE TWO RULES OF RETAILING...RESPOND TO THE CUSTOMERS NEEDS
SLIDE 17	GOOD - PROTECTING PRODUCTS VERSUS COST
SLIDE 18	GOOD - CONVENIENCE/TIME SAVINGS VERSUS COST
SLIDE 19	GOOD - DELI & BAKERY VERSUS LABOR/CONVENIENCE

SLIDE 20   BAD - RECYCLABLE WHERE? TRUCKLOAD
            QUANTITY

SLIDE 21   YOU'VE GOT TO SELL THE STUFF TO SOMEONE.
            COLLECTING IS NOT RECYCLING

SLIDE 22   ENVIRONMENTAL COMMITMENT - SUPER VALU
            WILL PROVIDE INFORMATION AS TO THE
            IMPORTANCE OF AND OPPORTUNITIES FOR OUR
            INTERNAL OPERATIONS, CONSUMERS, AFFILIATED
            OFFICIALS TO REDUCE, REUSE AND RECYCLE
            APPROPRIATE MATERIALS WHENEVER POSSIBLE.
            THESE EFFORTS ARE INTENDED TO REDUCE BOTH
            THE AMOUNT OF WASTE GENERATED, AS WELL AS
            ANY NEGATIVE IMPACT OF OUR BUSINESS
            ACTIVITIES ON THE ENVIRONMENT.

SLIDE 23   LETTER TO STRETCH WRAP SUPPLIERS - STRETCH
            WRAP IS BENEFICIAL TO OUR OPERATION IN
            REDUCING DAMAGE, AND CONSISTENT WITH OUR
            ENVIRONMENTAL COMMITMENT. IT IS IMPORTANT
            FOR OUR COMPANY TO FIND A WAY TO RECYCLE
            THE STRETCH WRAP USED IN OUR OPERATIONS.

            WHILE WE CAN TAKE PRIMARY RESPONSIBILITY
            FOR GATHERING THE STRETCH WRAP AT RETAIL,
            DIFFERENT VENDORS' PROGRAMS TO AID IN THE
            COLLECTION AND OFFSET THE LABOR COSTS WILL
            BE AN IMPORTANT PART OF OUR VENDOR
            SELECTION PROCESS.

            IT IS OUR INTENTION AND HOPE THAT BY YOUR
            WORKING WITH YOUR MANUFACTURER PARTNER,
            YOU WILL BE ABLE TO OFFER US A RECYCLING
            PROGRAM CONSISTENT WITH OUR NEEDS AT
            WHOLESALE AND RETAIL

SLIDE 24   OPPORTUNITY - THE PLASTIC SUPPLIER WHO ADDRESSES TRASH BETTER WILL HAVE A COMPETITIVE ADVANTAGE.

SLIDE 25   OPPORTUNITY - COMPETITIVE ADVANTAGE

...SIGNIFICANT MEASURABLE ECONOMIC ADVANTAGE OVER THE COMPETITION, AND IS PRESERVABLE AT LEAST IN THE SHORT TERM.

...NON-PRICE COMPETITION - CAN DIFFERENTIATE A "COMMODITY PRODUCT"

SLIDE 26   IN SUMMATION - GOOD - PLASTIC HAS MANY ATTRIBUTES

SLIDE 27   IN SUMMATION - GOOD: PLASTIC HAS MANY ATTRIBUTES - BAD: ATTITUDES AND PERCEPTIONS ARE IMPORTANT. INDUSTRY EFFORTS ARE FALLING SHORT OF PUBLIC EXPECTATIONS.

SLIDE 28   IN SUMMATION - GOOD: PLASTIC HAS MANY ATTRIBUTES - BAD:ATTITUDES AND PERCEPTIONS ARE IMPORTANT, INDUSTRY EFFORTS ARE FALLING SHORT OF PUBLIC EXPECTATIONS.

OPPORTUNITY - THE COMPANY WHO ADDRESSES THE "BAD" BETTER WILL HAVE A COMPETITIVE ADVANTAGE

THOUGHTS ABOUT PLASTICS PACKAGING FROM A FOOD RETAILER

William T. Boehm
Vice President
Grocery Procurement
The Kroger Co.
P.O. Box 1199
Cincinnati, OH 45201

Biography

William T. Boehm is Vice President, Grocery Procurement for the Kroger Co. He is responsible for purchasing the Company's private label grocery products, the commodities for Kroger's food processing plants and all private label packaging. He directs Kroger's investment buying operations. Kroger, one of the nation's largest retailers, has annual sales of $19 billion.

Boehm has a B.S. from the University of Wisconsin-River Falls and an M.S. and a PhD from Purdue University. He was Assistant Professor of Agricultural Economics at Virginia Polytechnic Institute before joining the U.S. Department of Agriculture in 1976. Before joining Kroger in 1981, Boehm was the Senior Economist for food and agriculture with the President's Council of Economic Advisors. Boehm is married and has two teenage sons.

Abstract

Plastics play an important role in today's consumer marketplace. Consumers like the convenience and lightweight afforded by plastic. Food manufacturers and retailers like its product integrity attributes. Its lightweight and non-breakable nature provides significant financial benefits as well.

Problems for plastic packaging loom, however. Legitimate concerns about the environment have put plastics packaging in an unfavorable light. Facts aside, plastics packaging is widely seen as an environmental polluter. Positive steps must be taken to change that perception. Speeches and consumer education programs will not be enough. Plastics packaging must be made easily recyclable, and soon!

"THOUGHTS ABOUT PLASTICS PACKAGING FROM A RETAILER"

Speech given by William T. Boehm, The Kroger Co., at the Plastics Institute of America Foodplas 1991 Conference in Orlando, Florida on March 5, 1991.

I've been asked to talk with you this morning on the topic: "THOUGHTS ABOUT PLASTICS PACKAGING FROM A RETAILER."

My initial instinct - from a person who buys $80 million worth of plastics packaging each year for our dairies, bakeries, beverage plants and grocery products plants was to compliment you on the product.

Plastics packaging has come a long way.

Plastics play an important role in today's marketplace.

-   Consumers like the convenience and lightweight
-   Manufacturing likes the product integrity attributes, the non-breakable nature.

You continue to make impressive progress on the shelf and in the home!

But, I'm not one to ever leave well enough alone! You don't need me to tell you that a lot of what you do is right!

You already know that --

What I want to do, instead, is share a very deep concern I have -- one that includes you and me in a very real way.

Fact is, ladies and gentlemen, it is impossible today to talk about packaging without talking about solid waste management -- plastics is a big part of that!

We are a throw-away society -- can't deny that fact. We have all heard the numbers. As a nation, we represent less than 7% of the population, but use more than 40% of the world's resources.

Numbers aside, and facts be damned, PLASTICS have come to represent the very worst that is in us!

Remember the Graduate? "Plastics, my boy, plastics!"

To complicate matters, we still discard our waste the old fashioned way! We put it in places where it won't be an eyesore.

At least, that's the goal!

We kid ourselves with fancy words like landfill in order to hide the real meaning.

When I was a kid growing up in Central Wisconsin, we called it a DUMP; that's what it was and that's what they are!

No one took any responsibility for the throw away part of our world.

In large measure, that's still the case.

At Kroger, we recently changed from glass to plastic for our private label peanut butter. A few people asked environmental questions.

But, it was very easy to rationalize - plastic or glass, it really makes little difference.

Both end up in landfills (dumps) today!

In fact, 80% of our garbage today is taken to a dump.

The problem now facing us is that there are more people (so there is more garbage), and it is getting more difficult to hide the dumps - NIMBY!

In 1970, we had some 15,000 landfills (dumps). Today, there are fewer than 5,000; by 1995, there will be fewer than 2,500.

EPA expects one in four cities will run out of landfill capacity by 1995.

Alternatives are not immediately obvious or easy.

Let me make four points:
1)   The waste problem is REAL!

    - It's not a problem created by activists or kooks.
    - It's one we've got to deal with.

2)   We are a major part of the problem and must, therefore, be a major part of the solution.

    Speeches, corporate policy statements, rationalizations by trade associations might make us all feel better, but they won't solve the problem.

    It makes no difference that plastics are only 7% of the solid waste stream, or that we've reduced the amount of PET resin used to make a 2 liter bottle.

    Those are the result of economic decisions - we are supposed to make those decisions as business people.

    I recently heard a speech by Ross Perot ..

    He made a point about the need for practical, focused ACTION!

    In a city with potholes? Get on the evening news, get a tear in your eye and express a deep concern!

    Sound familiar? Ross Perot says...

    "The world cries out for people who will get hot asphalt to fill a hole; then go on to the next one and do it without a press conference!"

3)   There is no simple solution to this problem -- For us to continue living on this planet, we must heed the advice of the environmental activities of the 1970s...

THIS IS THE GOOD SHIP EARTH

- There is no reason to argue about or even discuss who is to blame - we are all to blame.

- There is no reason to rationalize that one particular waste management approach is better than another - all approaches are needed.

4)   The solution is tied to both Culture and Economics

After all, the economic system is a reflection of the culture. Adam Smith wrote his wonderful book Wealth of Nations in 1776.

John M. Keynes published his treatise rationalizing a role for government and deficit spending in 1933, the very depth of the depression!

Get the connection?

Economic incentives/or the lack of disincentives cause people to behave in certain ways - change the incentives and you will change behavior! Aluminum cans are recycled today because there has been an economic benefit for doing so, that economic benefit created the infrastructure.

As the culture changes, it is possible to reinforce the economic incentive, but the reverse is also true.

Kroger is responding to this need for both cultural and economic change.

That should not be surprising - we are a part of some 1,200 local communities around the country. As a retailer, we see, every day, that packaging makes up a very visible 30+% of the municipal solid waste stream.

Without a lot of fanfare or speeches, we have made a commitment to be part of the solution. We do not have a corporate policy on this matter, we have an expectation to fulfill -- one that comes from within each of us, but one that is strongly reinforced by our senior management team, both at Corporate Headquarters and in the field.

Our commitment is to all three legs of the waste management stool:

1)   Source Reduction
2)   Recycling
3)   Consumer Education

1. SOURCE REDUCTION. Source reduction ought to be easy -- it makes economic sense to reduce the amount of packaging used in the food chain.

    But, it is never that easy, is it? Often we do things like double wrap packages because that's the way we do things (culture?). Marketing people don't often like to risk changing things that have worked. The Sudafed situation and others like it just reinforce our collective vulnerability.

    After working on this one for a while, I'm convinced that real progress will require significant effort from all of us.

    You want to sell packaging, plastics packaging. It's in your best interest to do so.

    The markets for used plastics are weak, at best!

    - Kroger manufactures some 2,000 private label products
    - We have committed ourselves to a systematic review of all packaging with an eye toward source reduction.

    So far, most of our progress in this area has been helped by economics. We have got to do better -- and we are working at it!

2. RECYCLING. We, like everyone else, realize that reuse has to be a big part of the overall solution.

    Every one of our retail divisions is deeply involved in a recycling activity using the retail store as the base point - plastics is a big part of that!

    Would we prefer curbside recycling in every community? Of course, but that's not being done and talking about it is typically just more hot air -- our motto...

        Find a pothole and fill it!

    Recent amendments that permit the re-use of recycled PET in food-grade packing is a major breakthrough for your industry.

    But, where is the infrastructure to collect the stuff?

    All must work together on the infrastructure.

    Without it, we have a lot of garbage collected and stored; with it we have a new natural resource.

The commitment we have made to recycling has had some significant benefits with our private label line.

- Our commitment is to use recycled paperboard for all folding cartons (except for HBC, we are there).

- To help simplify the recycling challenge, we shifted to all aluminum cans for Big K.

- Currently working with major resin suppliers, such as Paxon, to close the loop on HDPE recycling.

As I mentioned, we are collecting solid waste in almost <u>every</u> KMA -- we are working on the infrastructure.

- Virtually all 1,250 stores take back plastic grocery bags (10%)

In 1990, we recycled through our stores...

350 million aluminum cans,
25 million bottles and jars (glass)
40 million HDPE milk jugs and PET beverage bottles
65 million pounds of newsprint
10 million pounds of old phone books

Significant reductions in landfill tonnage; more important, we think we are leading the charge towards a cultural change.

Leads me to the final point...

3.   <u>CONSUMER EDUCATION</u>

Easy to interpret as rationalization!   Not the case!

Our position -- if it doesn't contribute to a solution of the problem, don't do it!

Kroger doesn't have <u>a</u> green line.   As a general statement, green lines are exploitive.   They are not directed towards problem solution! Biodegradeability for plastic bags is an empty promise.

Trees are <u>not</u> the problem; landfills are!

Don't get me wrong, our stores carry <u>green</u> <u>products</u>.   If consumers want to buy them, we will carry them.   But our <u>commitment</u> is to awareness, understanding and problem solution.   That's what explains our involvement in in-store recycling programs -- and in our aggressive office-level recycling activity.

Together, we will win this battle.

But we have got to stop rationalizing - we have got to start filling those potholes!

Started with a complaint -- and will end with one.

We have come a long way with <u>plastics</u> -- in spite of Dustin Hoffman -- but we have got a big job yet to do!

PLASTICS FOR FRESH, CHILLED, AND BAKED PRODUCTS

Judy Ordens
Director of Consumer Affairs
The Copps Corporation
2828 Wayne St.
Stevens Pt., WI 54481

Biography

Judy Ordens, Director of Consumer Affairs, The Copps Corporation, formerly worked as a research and development home economist in the Food Evaluation Center at Henri's Food Products, Milwaukee, WI

Education - Bachelor & Master of Science in home economics at University of WI - Stout. Professional Affiliations - Certified Home Economist, American Home Economists in Business, American Home Economics Assoc., Society of Consumer Affairs Professionals, Food Marketing Institute Consumer Affairs Council, Phi Upsilon Omicron, Portage County Wellness Commission

Abstract

As a small Wisconsin supermarket retailer and wholesaler, Copps operates in a fast paced service industry that is customer driven. Perception is reality in our business. For the most part, our merchandising and operations staff is happy with plastic packaging and overwrap because of its cost savings, convenience, sanitation, and safety features. But a major consumer issue is solid waste with reduce, reuse, and recycle as part of the solution. At this time, plastic is perceived as a large part of the problem and we may be forced to change direction and find other solutions for packaging.

PLASTICS FOR FRESH, CHILLED, & BAKED PRODUCTS

by Judy Ordens, M.S., C.H.E.
Director of Consumer Affairs
The Copps Corporation
Stevens Point, Wisconsin

Good Morning! I hope by the end of my talk you'll know something about plastics for fresh, chilled and baked products from a supermarket perspective. But more than that, I hope you will have a better understanding of our industry so we can better meet the needs of our customers. The supermarket industry is very fast paced and what's true today may be different tomorrow. I never thought I'd have to include a disclaimer in my speech, but as I began writing this talk I realized things were changing so fast that some of what I was writing in December would be outdated or changed by March. As unpleasant as that may sound, if you can accept that fact, you'll better understand our perspective of the supermarket industry. Consequently, I may deviate slightly from my speech in order to give you the most recent information.

This morning I will cover the following areas:

1. A brief overview of The Copps Corporation and the supermarket industry

2. Plastics in the Copps retail/wholesale business, especially in our produce, meat, bakery and deli operation.

3. Copps special programs involving plastics. This will include our "Make Everyday Earthday" campaign and Wisconsin's new recycling law.

4. I'll conclude with a summary of some of the important issues in our industry that impact on the plastics industry.

First I'll tell you about The Copps Corporation. Copps is a small family-owned company located in central Wisconsin in the city of Stevens Point. Copps was founded in the late 1800's by E.M. Copps. He was the great, grandfather of the current family members Mike, Tim, Tom, Fred, and Don who are active in the business today. Copps is unique in that it is both a retailer and a wholesaler. Copps owns and operates 16 corporate stores ranging in size from 15,000 to 85,000 square feet with weekly customer counts ranging from 5,000 to 32,000. The Copps wholesale operation supplies our corporate stores and another 44 plus independents. Stores are located in Wisconsin with our grocery and perishable warehouses located in Stevens Point.

Copps has been a member of the Independent Grocers Alliance, better known as IGA, since 1944. Most, but not all of our stores carry IGA private label products. As an industry, the primary role of a supermarket is to bring food producers and consumers together. Key costs in our business are for products, capital, labor, taxes, energy, and supplies. Food costs or "the food marketing bill measures total processing and distribution costs for American farm products from the farm gate to the consumer. Except for meat, produce, baked goods and other items packed in the store, the packaging of food products is done by the food manufacturer at the factory. In-store packaging, such as wraps, trays, and bags represents a small portion of the total packaging bill. Packaging accounted for 8.5% of the total amount the consumer spent on food in 1989." (Key Costs in the Supermarket Industry, 1990 - Food Marketing Institute)

The important thing to remember is that our industry is customer driven. Customer satisfaction keeps us in business. In order to keep up with the pace of retailing, we must react instantaneously and change constantly. As an example: if a competitor puts a low price on an item, we may match it or offer a similar promotion within a matter of hours.

Now I want to tell you about plastic as it relates to our retail and wholesale business. I'll begin with our wholesale distribution perspective and continue with grocery, meat, produce, bakery and deli merchandising at retail.

From a distribution standpoint, many plastic containers are not strong enough, consequently they don't hold up in distribution and storage at the warehouse level. Poor quality plastic rips and crushes, which causes excessive losses from product damage and cube usage. When I talk about cube, I'm referring to space used by a product. As an example: plastic drinking water jugs are a problem and special racking is needed because plastics are not self-supporting. Shrink wrap on the other hand is a positive. It helps prevent damage and is designed to keep product flowing through the system with the least amount of damage. In particular, I'm referring to the plastic wrap that is put around a pallet to stabilize product cases before shipping to a store. One of the biggest issues we face is recycling. Our costs are based on weight, cube, and labor. Recycling increases handling costs and may raise the cost of food. I'll go into more of this later when I tell you about Copps "Make Every Day Earthday" campaign. From our warehouse, I want to take you to our retail operation where plastic has much more visibility.

I'll start with store operations and grocery. At the checkout you'll hear, "Would you like paper or plastic?" Most of our stores offer paper and plastic bags. Plastic bags are cheaper, take up less space, and our competitors have them. We offer both so our customers have a choice. At our stores, customers don't bag their own groceries so productivity is a labor factor for us. It takes more time to bag plastic and it holds fewer groceries than paper. We do feel, however, that plastic bags are here to stay. We get customer comments such as: "Paper fits in my garbage can." - "I like the handles on the plastic bags." - "It's not degradable." Let me clarify. We consider degradability to be a non-issue, but the consumer still thinks paper is best because it degrades. Plastic has gotten a bad rap as far as solid waste and the environment are concerned. We know the facts about paper - it costs more, weighs more, takes up more space and doesn't degrade in landfills. But public perception says "no plastic." And we do need to admit that there are some terrible packaging examples where the quantity of product versus the amount of packaging is a joke. While plastic milk jugs have helped us to almost eliminate leakers, many containers don't make efficient use of cube. A box or cube is efficient; not bottles. Our grocery director says "think shelf space for distribution and retail. He also mentioned that wide lids for containers are better than small lids." Everyone I talked to at Copps said that recycling was a top priority for plastic. Think aluminum! Our grocery director said the only disadvantage to aluminum was that it wasn't clear. He'd buy aluminum grocery bags if they were available. He'd like plastic to be as good for us as aluminum. Get away from multi-layer plastics and produce simpler packages. Most consumers think all plastic is the same. Common resins for all containers would be idea. Plastic needs more value and it isn't crushable at home like the aluminum can. How about a plastic chipper? There needs to be more ground level recycling. Now, I know all of that is a tall order and some things aren't possible, but I'm confident there are workable solutions.

Now on to our meat department. Our meat director is pleased with the performance of foam meat trays. Pulp is more expensive and causes more sanitation problems. The soaker pads we use are a combination plastic and pulp. This past year, we did a test in a few stores to see if a soaker pad with recycled fibers would work. The idea was rejected because of performance and appearance, but in this business the idea may be revived tomorrow. We've looked at plastic packaging for fresh meat entrees and found acceptable trays and covers. There's good variety in microwavable trays and lids. Leakage is the major problem with any moist product. Even if the entire package is wrapped in plastic and sealed, it often leaks. We're solving this problem by using recipes with very little moisture. One problem we encountered with the label, not the package, was the fact that it melted in the microwave. And because of the adhesive it couldn't be removed before cooking. We have found labels that do work, however. Another factor in developing a meat entree program is the clear plastic lid that often pops off during microwaving. But we've found a solution to that, too. Customers view all plastic packaging as the same. They don't realize that some is microwavable and some not. Just like the electric coffee maker that says not immersible; plastic packaging must be permanently labeled for the consumer. They need to know how to use it and how to dispose of it. I've mentioned microwave! Microwave! But what about conventional ovens? I may be wrong, but when consumers want convenience they want microwavable, period! Our meat director feels the plastics industry needs to tell people about plastic, especially about recycling. Be more proactive or we may be forced to go back to pulp.

On to our produce department. Again, everything I say is subject to change - depending on consumer perception and the laws. Our stores offer a lot of bulk produce so we can get by with fewer containers. The containers we use are the same ones we use in the bakery. In our business, we try to turn or sell as soon as possible. That goes for all product, but especially in all fresh food departments. The plastic containers we use are better than pulp and have helped reduce product damage at retail. We like the clear containers so the customers can see the product. With pulp, customers think spoiled items are at the bottom. So with clear plastic we have a better level of confidence with the consumer. Shelf life on produce has improved. With the holes in the clear lids, there's less moisture residue than with pulp. Clear plastic has helped sell produce and it's helped with labor because our employees can see bad product when checking produce. It's more sanitary than pulp, can be reused, and is not damaged by moisture. But again the main issue is solid waste and we may be forced to go back to pulp. Disposal is the issue. We have paper ripening bags in produce and many rolls of large plastic bags. People keep asking for paper bags. As an aside, our plastic bags are the extra large so customers can easily put a large head of leaf lettuce in a bag.

In our bakery and deli, the clear snap on lid has helped with employee safety. Plastic overwrap takes more time and can be a safety hazard - from the heat sealing machine to the ergonomics in wrapping a package. Again the clear lids show off the product and help sell it. We do use some wax bags and cardboard boxes in bakery. Generally meat, produce, bakery and deli departments are very satisfied with plastic, but they are worried. Everyone I talked to is prepared to switch from plastic to pulp if it's necessary.

Next I want to tell you about the Copps "Make Every Day Earthday" program. Let me tell you that before we did anything we studied the issues. My files are jammed with reading material. All of which I've read during the past year and a half. We didn't come out with a bang on Earthday 1990. In fact, we developed our program in stages. Beginning with a pilot program in one store and gradually adding all 16 corporate stores and interested independents. There are many reasons for initiating a program such as ours. One is certainly the new Wisconsin recycling law or WI Act 335. Because the new law is so comprehensive, other states may follow Wisconsins' lead on the environment.

So let me review some of the pertinent aspects of the law as it relates to supermarkets.

1.   Wisconsin's new recycling law, signed by Gov. Tommy Thompson on April 27, 1990, is a vast, all-encompassing statute that will change the state's throw-away habits.

The law takes an innovative, phased-in approach featuring financial aids and technical assistance aimed at helping local communities start and/or expand programs that reduce, reuse and recycle wastes. The goal is to cut down on the more than six million tons of trash now going into Wisconsin landfills and incinerators each year.

Every home, apartment building, hospital, school, university, office, industry and governing unit - actually, every person and institution in Wisconsin - will be getting involved.

To assure widespread grassroots participation, the statute calls for statewide programs for recycling information and education.

2.   There are new solid waste management priorities (from most to least desirable):
   a.   waste reduction
   b.   reuse
   c.   recycling
   d.   composting
   e.   incineration with energy recovery
   f.   land disposal
   g.   incineration without energy recovery

Wisconsin Act 335 preempts local laws dealing with difficult-to-recycle packaging.

3.   What is a supermarket operator to do? One requirement is that owners of commercial, retail, industrial and governmental facilities either take their wastes to separation facilities or encourage occupants to recycle wastes. Those choosing the latter option must provide containers and transport the sorted waste to a recycling facility.

4.   What about products and containers?
   Product regulations - Truth-In-Labeling:
   The Dept. of Agriculture, Trade & Consumer Protection(DATCP) will establish standards for products which are advertised or labeled as being recycled, recyclable or degradable. The standards shall be consistent with nationwide industry standards.

   Plastic Containers:
   Plastic containers must consist of at least 10% recycled or remanufactured materials beginning January 1, 1995.

   Heavy Metal Content:
   Manufacturers and distributors must phase in prohibitions on selling packages, packaging materials or packaging components with specified concentrations of lead, cadmium, mercury and hexavalent chromium.

29

5. And When? Beginning January 1, 1995, the following items may not be put in a landfill, converted to fuel, or incinerated:

    1. aluminum containers
    2. corrugated paper and other container board
    3. foam polystyrene (in pieces and in molds used as protective packaging in shipping containers and in cups & plates used for serving food or beverages
    4. glass containers
    5. magazines and other material printed on similar paper
    6. newspapers and other material printed on newsprint
    7. office paper
    8. plastic containers
    9. steel containers
    10. waste tires (except when converting to fuel or burning to recover energy)
    11. bi-metal steel/aluminum containers for carbonated and malt beverages (above 5 points taken from Wisc. Dept. of Natural Resouces PUBL-IE-041 rev 6/90)

It's not perfect, but Act 335 is a law we feel we can live with. And more important than the law is what our customers are telling us. We keep in close contact with our customers soliciting responses from them on their shopping experiences. We conduct about 30 dinner meetings yearly with groups of customers. The environment has been brought up by customers at every meeting for the past 2 years. And each month we receive over 275 written comments from our customers. Here are 2 examples received last November. The first one is from a customer in a city with a population of nearly 100,000: "I came into your store looking for muffin cups to use in baking. Having always bought them in a paper container I was very surprized (in this day and age when we are becoming very concerned about non-biodegradable items) to find that the only ones you had in your huge store were Reynolds brand and they were in a hard plastic container which really has no other use but to throw away. I hope you will think of the environment when you choose your products."
The next comment came from a woman who lives in a rural area: "Speaking of improvement, have you been thinking of being a leader in recycling? You know that the grocery store is where people pick up 80% of all recyclable items; and the grocery store is the perfect place to return those items. Maybe a different recyclable item each week or each month, whichever could be arranged by you."
When the Copps environment committee began meeting over a year ago, we did a lot of reading and agonizing on how to attack this issue. We don't want to be a recycling center. Our program was not to be a public relations stunt for Earthday. Some of our competitors advertised programs with full-page newspaper ads, but when we went to the store nothing was available such as brochures or canvas bags and employees

didn't know anything about their program. We decided to wait, plan carefully and begin with a pilot program in one store beginning September 1990. Customers and employees were very enthusiastic so in December we introduced the program to the Copps Food Centers. And in February, all of our IGA stores, plus our independents signed on. As of now _____ stores participate in the "Make Everyday Earthday" campaign.

Let me tell you some of the components:

1. At Copps, we:

   - offer reusable canvas grocery bags for sale to customers. We pay customers $.05 per canvas bag used on their orders.

   - pay customers $.05 per paper bag reused on their orders.

   - use paper bags made from 35% - 100% recycled material

   - provide a bin in the store so customers can return their plastic bags for recycling

   - established a "cans for cash" program in the parking lot for customers to deposit aluminum beverage cans for recycling.

   - display and sell cloth diapers as well as disposables

   - eliminated all foam egg cartons

   - stock "Green Forest" products

   - use recycled paper in our advertisements, flyers, and mailers when available

   - use of refrigerant with the lowest levels of CFC

   - put a can recycling program in place for store employees

   - distribute informational material to our customers and employees to encourage individual participation in our local recycling program

   - talk with consumer board members for ideas on how to better serve our local environmental needs.

2. Copps is currently recycling 500,000 pounds of fat, bones, and grease annually. Over 6 million pounds of paperboard cartons are also recycled each year, as are aluminum cans. Old bakery products are given to organizations for use as animal feed. Copps energy-efficient system recycles cold and hot air from refrigeration and compressors to save natural gas.

3. We encourage our stores to work closely with their municipalities to promote local recycling programs, especially curbside pickup. Our warehouse offices and some stores participate in local office paper recycling programs. In addition, our corporate stores collect shrink wrap and some other plastic which is shipped back to our warehouse for recycling. This last component and the consumer plastic bag recycling causes the most problem. Clean, empty, dry plastic bags and a recycler who will take them are issues we wrestle with constantly. Customer and employer acceptance has been overwhelming. Interestingly, our younger employees are the most enthusiastic. And for a supermarket that's wonderful because so many of our employees are teenagers. Between 60-70% of the workforce at the store level is part-time, so 16-19 year olds comprise a large number of our employees.
Now I'd like to show you a video of some clips from the Copps "Make Everyday Earthday" program. One of the first steps was to develop a logo which appears on our brochures, ceramic cups and canvas bags. This is a television commercial developed for our first test of the program in September. The kick-off promotion centered around the Muscular Dystrophy Association carnival/telethon over Labor Day weekend. Proceeds from the sale of the canvas bags were donated to the MDA. For the kick-off in the Copps Food Centers we built the promotion into our 12 days of Christmas event. The audio is a radio commercial. We got wonderful newspaper and TV coverage. And as I've said, we've had wonderful customer acceptance. Here are some scenes at the checkouts of canvas, paper, and plastic bags. Customers have overwhelmed us by bringing back many plastic bags for recycling.

I want to reiterate that our program is on-going so we are always looking for new things to do. We have shied away from labeling certain products as environmentally friendly or bad for the environment. But, several times employees have taken it upon themselves to help. This is the first unauthorized sign that was posted. Here is the next.

Office paper recycling seems to be working. Our biggest problem is finding a recycler who will take more of our plastic. Here's a photo that I'm sure you'll find offensive, but it was a great success for us. We introduced the Copps Earthday cup to our office employees. Styrofoam is the bad guy once again only because that's what we buy. Paper cups are 2-1/2 times more expensive. Interestingly some people, such as our V.P. of Finance, are savvy enough to understand that washing cups takes time, hot water, detergent, and towels.
Enough about Copps. It's the consumer that really matters. What do they want? First they want easy to read accurate information, not public relations material. They're tired of plastic bashing and paper bashing. Recyclable doesn't mean much if

it still goes into a landfill. "No material is truly recycled until it is bought back into productive use in manufacturing and production." (University of WI - Cooperative Extension newsletter 11-90, "Working for Wisconsin Families".)

I've spent a lot of time today on environmental issues because that is where the most urgent work needs to be done. But there are other consumer issues that I feel are important. And remember, a consumer issue is our issue. Customers want heat resistant, microwavable, dishwasher-safe containers. Melamine, cottage cheese containers and margarine containers are still being used in microwave ovens. I'm personally concerned that toxins from plastics used in microwave ovens may cause real problems in the near future.

In "Future Food for Americans," Dr. John Stanton talks about the new "time-starved" consumers and "dashboard dining." He continues, "95% of all households will have at least one microwave oven by 2001. ... The microwave niche continues to expand, particularly shelf-stable items with non-metallic retort packaging and refrigerated microwavables which are chilled, not frozen. The forecast: shelf-stable retort or tub products could capture a full 50% of the canned food market and reach sales of $3.6 billion. Combined with sales of refrigerated microwavable products, sales could reach $5 billion." (Food Nutrition News - Nat'l Livestock & Meat Board vol. 62 no. 5) I could go on and on with information concerning children, minorities and the senior citizen. We face these issues everyday and we're trying to meet the needs of our customers with more fresh, healthy easy-to-serve foods from our meat, produce, bakery, an deli departments. And, of course, plastic is essential for packaging.

In conclusion, our merchandising staff, for the most part, is happy with plastic packaging and overwrap because of its cost savings, convenience, sanitation and safety features. But a major consumer issue is solid waste with reduce, reuse and recyle as part of the solution. People think plastic is toxic to burn and it doesn't degrade. Perception is reality. At this time, plastic is perceived as a large part of the problem and Copps may be forced to change direction and find other solutions to plastic.

I quote from Tim Simmons of Supermarket News, November 12, 1990, "If business and government leaders interpret the cooling of media interest as a reason to reduce their commitment to environmental programs, it will be a painful and costly mistake."

PLASTICS PACKAGING AND THE ENVIRONMENT:
YOUR FACTS, OUR FACTS, REALITY

Michael Reilly
Manager, Environmental Affairs
Wakefern Food Corporation
600 York St.
Elizabeth, NJ 07207

Biography

Michael Reilly is Manager of Environmental Affairs, Wakefern Food Corporation, Elizabeth, New Jersey. Previous to this he was a Manager in the company's Human Resource Division responsible for retail recruiting as well as developing community projects that improve the reputation of Wakefern/ShopRite Supermarkets in the communities they serve. He started in the Food Industry in 1980 and has held a number of positions which included Night Crew Supervisor, Department Manager, Corporate Trainer as well as a Regional Personnel Supervisor.

Abstract

Our presentation will take a look at the environmental claims that plastic packaging companies are making through public relation/education campaigns. How these claims are received and understood by wholesalers, retailers, consumers, environmentalists and legislators. The last part of the presentation will look at proposed environmental packaging guidelines and legislation which includes plastic packaging bans.

FOODPLAS 91

PLASTICS PACKAGING, YOUR FACTS, OUR FACTS

REALITY

PRESENTED BY MICHAEL REILLY, MGR.ENVIRONMENTAL AFFAIRS

WAKEFERN FOOD CORPORATION

It is a pleasure to be here with you this morning. I am the manager of environmental affairs for Wakefern Food Corporation/ ShopRite Supermarkets. Wakefern/ShopRite is the largest retail owned cooperative in the United States. We operate 180 supermarkets in Connecticut, Delaware, Massachusetts, New Jersey, New York, and Pennsylvania. We serve over three million customers each week.

I would like to thank Mel Druin for giving me the opportunity to be a part of this conference. When we first spoke, he mentioned that there were members of the Plastics Institute of America (P.I.A) who felt that they were not always getting a full and accurate picture of how plastics packaging is viewed by wholesalers, retailers, and consumers in regard to the environment. I thought it might be beneficial if you were told about some of the experiences Wakefern has had over the last few months regarding plastics.

My presentation will cover the following topics:

- HOW RETAILERS,WHOLESALERS,CONSUMERS VIEW

  PLASTICS ENVIRONMENTAL INFORMATION
- LEGISLATIVE INITIATIVES
- TOWN MEETINGS
- SOLID WASTE MANAGEMENT TRENDS
- WORKING TOGETHER FOR A POSITIVE CHANGE

As WFC puts its environmental plan together, we continue to take the position that we will not get involved in or promote any good product/ bad product campaigns. We will, however, provide factual environmental information to help our customers make informed purchases. Due to the complexity of environmental issues, most companies are taking this approach. Plastics companies are no exception. In a White paper that Dow recently published, there was an excerpt that stated the following, "There's no shortage of environmental news in both the trade press and the consumer media. But when it comes to reports concerning the relationship between plastics and the environment, an alarming percentage of that "news" is inaccurate, misleading, or factually distorted." We agree with this statement. Our experience shows us that some plastics companies are the worst offenders when it comes to spreading the previously mentioned "news".

The continuing debate on plastic versus paper grocery bags is a good illustration of this. Some of you may be familiar with the Resource and Environmental Profile Analysis of Polyethylene and Unbleached Paper Grocery Sacks study. This study was completed by Franklin Associates for the Council for Solid Waste Solutions. I was introduced to the report when a customer who worked for a large New Jersey plastics company sent me the first chapter. It was his response to a bag reuse program we had just introduced in our stores. He felt that we unfairly positioned plastic grocery bags in a brochure we distributed to our customers. It had a question and answer format. He took exception with the following two questions.

Question 2: Should I reuse plastic bags? paper bags? both?
Answer: All should be reused if possible. Reusing paper bags reduces the number of trees that must be harvested. However, trees are a renewable resource. The petroleum used to make plastic is limited and unrenewable so plastic bags should be reused also.

Question 3: How will my reusing of bags help save the environment?
Answer: By reducing the number of new bags manufactured and in circulation, you will conserve the number of trees (paper bags) or amount of petroleum (plastic bags) used. You will also reduce the amount of waste filling our already crowded landfills.

His letter gave the impression that the Franklin study definitively showed that plastic grocery bags were more environmentally friendly than paper bags in regard to atmospheric, water, and solid waste. After reviewing the data, I had more questions than answers. For example, the study showed that ten thousand plastic bags created a little more than one pound of waterborne waste compared to over thirty pounds of waterborne waste for the equivalent number of paper bags. There was no mention of what the waste was comprised of or what negative impact these wastes may have on the environment.

One of my concerns with this report is that it falls short of its goals. Franklin states,in Chapter 1, "the purpose of this study was to determine the energy and environmental impacts of polyethylene and paper grocery sacks." In Chapter 2,it states the following: "The environmental impacts include wastes in the form of air emissions, solid waste, and waterborne waste. No attempt is made to describe what the effects from these wastes may be. In other words, no attempt has been made to determine the relative environmental effects of these pollutants such as fish kills or groundwater contamination as there are no accurate data available." I spoke to a number of other environmental managers as well as scientists in and outside the food industry. They had similar concerns with how the report was written and distributed. Apparently, the study was being used as a public relations and marketing piece by some companies. Unfortunately some of the data presented in the study has been used for materials bashing which really helps no one.

We could probably take the rest of today discussing the merits of environmental impact studies and their use. We could also debate the pros and cons of one package versus another. But as we were enjoying these discussions I'm afraid that there would be another piece of legislation either proposed or passed that will impact our businesses. In New Jersey, this past legislative session opened with over one thousand proposed pieces of environmental legislation. These bills range from prison terms for executives whose companies accidentally violate environmental laws to one that will fine any retailer who carries packaging that does not meet state guidelines. At the local level we have over 37 communities who have introduced or passed plastics packaging bans in their communities. The supermarket chains in our market area have continued to fight this type of legislation. However, it's becoming more difficult. With more factual information we can better present our case. But who's facts do we use? Biodegradability is the issue that most communities use as the reason for passing plastic bans. Normally, we try to explain that the majority of our waste is land-filled and nothing really degrades in that environment. We've attended public town hall meetings where representatives of the plastics industry have argued that because nothing really degrades in a landfill that plastics are better because by weight they are the smallest percentage of landfill waste. We're seeing more opponents to plastics presenting their own data which shows plastics taking up 18 percent of a landfill by volume. To the credit of the representatives of the Council for Solid Waste Solutions they have consistently used both weight and volume in presenting plastics impact on landfills. This presentation of the total picture can only be a benefit to your industry.

There has been a major shift in the New Jersey solid waste management goals. Currently we are recycling approximately 36 percent of our solid waste. There were 15 incinerators, site approved, to be constructed in the state. With a further increase in recycling and these new facilities, New Jersey probably was on its way to managing its municipal solid waste within its own borders. There is currently a moratorium on incinerators and a new emphasis being placed on recycling. The state has a goal of 60 percent by 1995. This can be a great opportunity for plastics recycling, not only for New Jersey but for the rest of the country as well. There are a number of states, legislators, and environmental groups who will be closely watching New Jersey to see if we are able to manage our municipal solid waste with a minimum of incinerators.

The next thing I would like to discuss is a meeting I recently had with a community recycling group who had requested we stop selling styrofoam cups and plates in our stores. This particular community is affluent and situated on the Jersey shore. Their primary concern was the amount of styrofoam that was washing up on their beaches and waterways. After a rather lengthy discussion, it was evident that this community had a serious litter problem. We argued the point that banning the sale of these styrofoam products would not eliminate the litter problem. We suggested a number of anti-litter programs that they might want to get involved with and offered our services in producing some educational pieces. These consumer brochures will focus on migratory birds, litter, and other environmental issues concerning the seashore. We are not sure whether the community will drop the idea of initiating a styrofoam ban. The interesting part of this meeting was finding out how people perceive the various issues and how and where they get their information. It was evident that these were very sincere and concerned individuals who were frustrated with the environmental problems of their community.

You may find interesting this following piece of information that was presented to me as one of the reasons that we should be willing to stop the sale of the products mentioned earlier. And I quote,

" <u>More Bad News On Polystyrene.</u>

As if it's not bad enough that Polystyrene (styrofoam) disposable products and packaging stay around forever in the environment, and the chlorofluorocarbons (CFC's) used as a "puffing" agent cause ozone layer depletion, and worker and environmental protection are at risk during their production, now the Foundation for Advancements in Science and Education has revealed that styrene monomer has been found in human fatty tissue. The probable source: use of styrene-based disposable cups and food packaging.

Styrene has been well documented to cause neurotoxic effects (fatigue, nervousness, difficulty sleeping), hematological effects (low platelet and hemoglobin values), cytogenetic effects (chromosomal and lymphatic abnormalities), and carcinogenic effects.

The industry's response to these health concerns is to develop new marketing strategies to reverse the public's negative opinion about plastics and styrofoam products. They have backed their promotional campaign with $150 million.

Let the staff at your supermarket, the corner deli and restaurants know you'd rather your food and beverages never get close to styrene."

This information was provided to the group by the New Jersey Environmental Federation and The Institute for Local Self-Reliance in Washington, D.C.

In conclusion, I would like to make the following comments and suggestions. Those of us who are responsible for helping direct our companies to be as environmentally sound as possible have a difficult task ahead of us. There are, however, some things that we can all do to make the task easier:

1. Pay special attention to informational materials your company may be using to educate your customers. Make sure that it presents the total picture.

2. All of your associates should take an active part in their communities' legislative process. This will ensure that when environmental issues are presented that communities pass legislation based on facts.

3. As an industry, there needs to be an acceleration of plastics recycling throughout the nation.

4. Work closely with your customers in the food industry to develop real packaging source reduction.

5. Stop materials bashing.

If all of us in the food industry take a pro-active role on these environmental issues we can take pride in the fact that we have made positive changes for the customers and communities we serve. Thank you.

SESSION 2

HOW FOOD PROCESSORS LOOK AT PLASTICS

TEAMING UP PLASTIC PACKAGING WITH FOOD PRODUCTS
A 10-YEAR RETROSPECTIVE & AN ASSESSMENT OF THE CHALLENGES AHEAD

Judy Rice
Senior Associate Editor
Food Processing Magazine
Putman Publishing Company
301 E. Erie Street
Chicago, IL 60611

Biography

After 7 years association with USDA's Food & Nutrition Service, Judy joined Putman Publishing Company in 1978. Over the past 13 years, as a member of the editorial staff of FOOD PROCESSING Magazine, she has reported on packaging developments for FP. She has a BA degree from Bowling Green State University, Bowling Green, Ohio and post graduate studies at Arapahoe College, Littleton, Colorado.

Abstract

A brief review of the progress of plastic packaging for foods over the last 10 years, plus an assessment of the current situation and future challenges facing food companies who use plastic packaging and supplier companies who manufacture plastic packaging.

TEAMING UP PLASTIC PACKAGING WITH FOOD PRODUCTS--A 10-Year
Retrospective & An Assessment of the Challenges Ahead

Judy Rice, Senior Associate Editor
FOOD PROCESSING MAGAZINE

It really is an honor to be here. FoodPlas year in, year
out has proven to be an outstanding packaging conference. As an
attendee, I've always learned something and met someone who could
be a good information source for a future article. And I want to
thank Mel Druin for inviting me and FOOD PROCESSING MAGAZINE to
participate from the podium this year.

I'm not a packaging engineer, nor a food technologist, nor a
plastics chemist, nor a microbiologist--nay, not even the chief
executive officer of a food company or packaging company. What I
am is a journalist. And in that capacity I don't really have to
know a lot about anything. I don't have to be an expert, but I
have to have contacts who are. So, as I look out over this
audience, I see a gold mine.

You are the people who have the inside information about new
products under development and about new food packaging
applications that are being targeted for commercial introduction
6 months, a year, 5 years down the road. You are the people who,
by being willing to share information and knowledge, have taught
me and many other food and packaging magazine editors pretty

close to everything we know. And, certainly, I still have a great deal to learn. So, as you leave the conference room today, if you all would just kindly drop one of your business cards in the fish bowl I've arranged to have placed outside the door, it would be greatly appreciated. I'll be in touch.

So, now, let's talk about what we're here for--a subject close to all our hearts--plastic packaging for food and beverage products. My talk has a very ambitious title: Teaming Up Plastic Packaging With Food Products--A 10-Year Retrospective and an Assessment of the Challenges Ahead. I hope I can live up to the expectations that title suggests.

Before we get into where we've been and where we're going with plastic food packaging, let's just take a quick look at where we are.

According to data gathered by our research department at Putman Publishing Company, 81.4% of the readers of FOOD PROCESSING Magazine use plastic containers for at least some of their products. Of course, many food companies have diverse product lines and offer various products in a range of packaging types. But overall, as this graph shows, plastics lead the pack. (SLIDE 1)

According to figures supplied to me by one of my trusty contacts at Du Pont, the market for all food and non-food, flexible and rigid plastic packaging is estimated at $24 billion or about 40 billion pounds. Of this $24 billion, 78% represents

<u>food and beverage</u> packaging. So that's an estimated $18.7 billion market for plastic food and beverage packaging.

The worldwide <u>percentages</u>, according to Du Pont, are 51% for liquid foods, 19% for snack food packaging, 7% for meat and cheese packaging, and 1% for convenience food items such as microwaveable products. Obviously, there looks to be a tremendous potential for growth in that last category.

And a word with regard to that new genre of food products-- refrigerated dinners, entrees, and salad items in modified atmosphere packaging. There simply is no packaging material so well-equipped to handle the requirements of these products as barrier plastics. (SLIDE 2) Last year, FOOD PROCESSING MAGAZINE did a sample survey of 1000 of our readers. One of the questions we asked: Do you expect to introduce any refrigerated products in modified atmosphere packaging in the 90s? 28% said yes.

If that 28% is representative of the food industry as a whole, and we think it is, then there is a major marketing opportunity out there for plastic packaging suppliers.

A 10-YEAR RETROSPECTIVE

Now let's do a little mental time travel. Let's all imagine we're 10 years younger. I'll know you're really concentrating when I see some silly grins begin creeping across your faces. It's 1981. So, let's think back. What was happening around that time? We-1-1-1-1, Ronald Reagan was in his first term as President. Chariots of Fire and My Dinner With Andre were playing at the box office. And I myself had just graduated from college—and, if you buy that, I can sell you anything today.

Now, in 1981, where were we in terms of plastic packaging for food products? Well, we had plastic film wraps for meats, cheeses, breads, and such. Most of that had been around since the 40s. And we had plastic film boil-in-bags (SLIDE 3)—around since the 50s. And retort pouches—shelf-stable foil pouches which incorporated a layer of plastic film—also around since the 50s (SLIDE 4)—though this particular product, the Kraft a la Carte, didn't debut until the late 70s. Aseptic portion pack creamer cups made from plastic—around since the 60s. And, of course, we had PET soft drink bottles. They came on the scene in the 70s.

So, by the time the decade of the 80s rolled around, those were still pretty much the basic categories of plastic packaging for food products. But things really started popping in the 80s. Boy, did they ever!

ASEPTIC PACKAGING

In January 1981, the Food & Drug Administration approved use of hydrogen peroxide as a package sterilant. That action opened the floodgates for a tidal wave of products, primarily juices, in aseptic carton packaging (SLIDE 5). Born of European technology, the cartons are constructed of 65% paper, 30% polyethylene, and 5% aluminum foil. The hydrogen peroxide is used to sterilize the interior of the packaging. Then the product (SLIDE 6), which has been heat-sterilized at a high temperature for a short time, and the sterile packaging are brought together in a sterile chamber, where product is filled, and package is sealed. The result is a high-quality packaged product which is shelf-stable for an extended period without refrigeration.

Now in case it's not immediately obvious how the emergence of the aseptic carton in the U.S. marketplace and around the world has impacted on the plastics industry, Du Pont--a major plastic resin supplier--refers to Tetra Pak--a major aseptic carton supplier--as one of its biggest global customers.

In the early 80s, aseptic juice boxes proliferated in the U.S. They were a marketplace phenomenon, and their success was so impressive that competing forms of aseptic packaging quickly made their way into American supermarkets. These new players included aseptic barrier plastic cups (SLIDE 7) and plastic bowls and aseptic multi-layer plastic film pouches (SLIDE 8) sterilized with hydrogen peroxide, steam, hot air, UV light, gamma irradiation, or combinations thereof.

And as the forms of aseptic containers diversified, so did the products contained in these packages. Juices still dominate the aseptic category, but other products (SLIDE 9) such as milks, tomato sauces, apple sauces, wines (SLIDE 10), puddings, cheese sauces, and soups (SLIDE 11) have dabbled in aseptic packaging with varying degrees of market success. Even tofu (SLIDE 12) has been marketed in aseptic packaging.

Particulate-containing low-acid foods still present a difficult challenge and an exciting opportunity for aseptics. It's more difficult to ensure all-the-way-through sterility of particulates such as meat chunks. And in form/fill/seal aseptic systems, container sealing is trickier. Pieces of food can get lodged in the seal area. However, the general sense is that these challenges will be met and satisfactorily dealt with by the end of this decade.

In both the consumer retail marketplace and foodservice markets, aseptic packaging--much of it employing plastic in some form as a packaging component--will grow, diversify, and thrive.

So, plastics in both flexible and rigid forms have been and will be key components of aseptic packaging. But aseptics were not the only food packaging technology to grow in the 80s through incorporation of plastics.

COEXTRUDED PLASTIC BARRIER PACKAGING

By the middle of the decade, we were hearing more and more about new coextrusion technologies (emanating primarily from Japan) that enabled the combining of an oxygen barrier plastic such as ethylene vinyl alcohol (EVOH) with more conventional packaging plastics such as polypropylene. The result was containers—such as condiment bottles (SLIDE 13)—which protected the product as well as glass did, but at the same time were lightweight and unbreakable (SLIDE 14). Consumers liked the idea, and food processors saw a major marketing opportunity. Several products traditionally packaged in glass were switched to coextruded oxygen-barrier bottles in the 80s.

RETORTABLE PLASTIC PACKAGING

During those same years, we also saw the evolution of "The Retortables"— (SLIDE 15) multi-layer plastic trays and bowls engineered to stand up under retort heat-processing conditions. In addition, we saw a new type of retort pouch—an all-plastic construction (no foil layer) designed for microwave heating.

Retort processing itself has been around since the 1800s. Simply stated, the process involves filling uncooked or partially cooked product into containers—traditionally metal cans or glass jars—hermetically sealing them, placing them into a retorting kettle, closing the retort, displacing internal kettle air with hot water or steam, and cooking at desired temperatures (in the neighborhood of 240 to 250 F, depending upon the product).

The retort process results in shelf stability. For example, a meat-containing product such as a can of chicken soup or a jar of beef stew needs no refrigeration and can be safely stored in the cupboard because of retort heat-processing. Particulates pose no great problem for retorting, primarily because heat is applied for a longer period of time than with aseptic product sterilization.

Over the years, metal cans and glass jars, and to a lesser degree foil/polypropylene pouches have been the mainstays of retort processing, because they provide an airtight seal to protect the product, and because they can withstand the high heat (240-250 F) required to ensure commercial "sterility" of finished shelf-stable product. But in the mid-80s, a new option came along--rigid plastics that can take the heat of retorting.

Typically, these new containers have been constructed of several layers--including polypropylene layers, adhesive layers, and a center oxygen-barrier layer of EVOH or PVDC. We've seen them in the marketplace in the form of plastic cans, bowls, and trays with metal pull-tops and plastic/foil peelable lidding.

Hormel has been the leader in commercialization of several products in retorted plastic packaging. Campbell Soup (SLIDE 16) has been extremely active in test marketing products in retort plastics. Other food companies in the vanguard of retortable plastic packaging include Dial Corp. with its Lunch Bucket and the recent Light Balance spin-off line; Del Monte with its Vegetable Classics; and last year Pet Inc. with its Progresso to Go line of healthy meals in retorted plastic trays. NutriSystems also uses retortable plastic containers (SLIDE 17).

Typical non-refrigerated shelf life for these various products is 18 months. Their appeal for consumers is first and foremost convenience. Frankly, as a consumer, I have to be honest and say there may be some organoleptic problems with some of these products. I'm a smoker--so I'm not supposed to be able to smell or taste anything. But in some of these products in retorted plastic containers, I have dectected flavor and odor transfer from the plastic to the food. So there's still a challenge there for packaging technologists and food scientists.

Still, while food processors and consumers certainly don't want to forsake quality, the appeal of these products is first and foremost convenience. You don't have to worry about refrigeration. You don't need a can opener. You don't need a pan to heat the product. You may need some eating utensils--a plastic fork will do nicely, thank you.
You can put a Hormel Top Shelf Beef Roast & Potato entree into your brief case and take it to work with you. Just pop the tray into the microwave oven at the office.

PLASTIC PACKAGING &
THE MICROWAVE OVEN

And that brings us to the subject that may be the single most powerful catalyst for growth of plastic food packaging--the microwave oven. Plastic packaging and microwave ovens just seem to go together. They are a natural fit.

In the latter part of the 80s, we began to see so many frozen products in microwaveable plastic containers (SLIDE 18). During that time, there was a surge of enthusiasm for CPET--crystallized polyester. CPET gives the consumer a choice. He can put the tray in the microwave oven or in the conventional oven. Not all plastic trays can handle this dual-oven application. Some paperboard and foil trays also present dual-oven problems. And in the 80s, many food companies felt they needed to provide this option to their customers. So CPET was hot.

Over the past 5 years, the attitude has changed. The situation has changed. Very soon nearly 90% of American households will own at least one microwave oven. The option of heating a packaged prepared meal in a conventional oven has become almost superfluous.

Add to that the practical matters of package cost (dual-ovenable generally costs the food processor more than microwave-only packaging), and of food formulation (it's very difficult to formulate a product that will look and taste the same regardless of whether it is heated in a microwave or conventional oven), and the combined result is a strong trend toward microwave-only packaging.

However, CPET does remain the material of choice for a

number of food processors. In 1990, for instance, ConAgra (SLIDE 19) chose to launch its new Kid Cuisine^{TM} frozen dinner line in CPET dual-ovenable trays.

Various plastic tray structures, along with paperboard, fiber, and coated foil trays, are vying for the frozen foods market. And it's going to be interesting to watch this competition. In very large measure, how well plastics do in this category, and, in fact, in most food and beverage categories, is going to depend on how the food and packaging industries confront the nagging question of "GREEN PACKAGING."

PLASTIC PACKAGING
& THE ENVIRONMENT

Be assured that the question of environmentally friendly packaging isn't going to go away. It's a theme that will dominate the 90s.

Many of you would argue (and the statistics support) that packaging is not the archvillain which has single-handedly robbed us of landfill space. Newspapers, phone books, and yard waste—all big contributors to the problem—just don't seem to take as severe a beating on this issue as does packaging in general and plastic packaging in particular.

However, food processors certainly realize that, as they try to market their products, consumer perception is nine-tenths of the law. Consumers want packaging convenience without guilt.

They want to be assured that a package recycles, or burns cleanly, or biodegrades, or otherwise disappears in an accommodating manner. And they expect food companies and packaging companies to deliver the goods.

As noted earlier, in March 1990, FOOD PROCESSING Magazine published a packaging survey, based on a sample of 1000 of our readers. (SLIDE 20) When asked how important recyclable packaging would be in the 90s, 62% said extremely or very important, 31% said somewhat important, and only 7% said not important at all.

In response to the question of how important will biodegradability be, 57% said extremely or very important, and 35% said somewhat important. I suspect that the percentages on the biodegradable question would be slightly different today than they were a year ago. Degradable packaging as a workable solution to the solid waste crisis has undergone severe scrutiny over the last year.

I'm sure that many of you get a little agitated, maybe even a little defensive when you hear people talk about "green packaging". Probably, you want to challenge them to define exactly what green packaging is--degradable, recyclable, incineratable? Labels by themselves don't mean a lot, after all. Nothing degrades in a landfill without proper exposure to air and water--two elements which many landfills don't have much of. And if these packages were to degrade, what sort of chemicals and toxins might they leach into the surrounding soil and water?

Numerous types of packaging can be incinerated, and the heat can be recovered as valuable energy--but safe incinerators are expensive to build and difficult to site. So what does it matter that packages can be burned, if there are not enough places to burn them, and so long as the public perception of incineration is so negative?

And with regard to recycling, we need to remind consumers that no package will recycle itself. There has to be consumer involvement, and there have to be collection and recycling networks in place, and there have to be markets for recycled materials and for products made from recycled materials.

A fair percentage of empty containers--regardless of what they are made from and regardless of whether they are virgin or recycled varieties--eventually will stray into the waste stream, despite stepped-up collection efforts. For all our good intentions and years of experience with recycling other container types, we've still got a lot of work to do on the collection front.

Let's look at the statistics. The aluminum industry recycles about 60% of its beverage cans. The Glass Packaging Institute estimates about a 30% recycling rate for glass containers. Steel cans overall (that is, food cans, beverage cans, and non-food and non-beverage cans) are being recycled at a rate of about 18%. The recycling rate for PET beverage bottles is approximately 20%. These stats tell us that substantial numbers of containers are getting away, and they're likely ending up in landfills.

So, again, what do these "green" labels mean after all? Well, they probably help sales, and they alleviate the guilt feelings of consumers, giving them moral permission to buy the packaging types that they really prefer in the first place.

For many, many products, consumers simply prefer plastic packaging. It's lightweight, unbreakable, often shelf-stable, recloseable, and microwaveable. It's convenient. And convenience is something today's consumers just can't get enough of. Plastic packaging fits the bill.

THE CHALLENGE OF CHANGING
PLASTIC'S ENVIRONMENTAL IMAGE

I don't want to spend a lot of time on the topic of plastic packaging and the environment, because several other speakers eminently more qualified to address the subject will be doing so during the course of this conference. But I do want to end on an optimistic note with respect to the outlook for enhancing the environmental image of plastic packaging. (SLIDE 21) Such image enhancement is going to be crucial to plastic packaging's ability to hold onto that 81.4% food usage rate—which was mentioned at the outset of this engrossing, hard-hitting, and eloquently delivered presentation. If we are going to talk about challenges for plastic food packaging, then revising the ecological perceptions certainly has to be the biggest.

It is inevitable that incineration eventually will gain wider acceptance as a waste management option in this country. And when that happens, plastics will hold an advantage. Plastics are abundantly BTU-rich. And in the face of growing concerns about energy costs and energy conservation, they represent a readily available energy resource just waiting to be tapped.

Technologies exist to enable environmentally safe incineration. They don't come cheap. But in the long run, return on investment could be handsome—both in dollars and in ecological benefits. One logistic difficulty that does need to be confronted with respect to incineration of packaging and other solid waste is disposal of the ash. Perhaps it is of necessity destined for a landfill, or perhaps some useful, imaginative, safe fate can be conjured up by resourceful minds.

While we wait for incineration to "catch fire" so to speak, efforts at collection and recycling of plastics will make headway. According to the EPA, only 2% of plastics currently is being recovered and recycled. So, there's a whole lot of room for progress, and the challenge is formidable. True, plastics in all their infinite varieties, carry recycling complications the likes of which the aluminum can industry and the glass packaging industry have never seen. But, necessity is the mother of invention.

New plastic technologies will emerge to help solve some of the current puzzles associated with the recycling of complex plastic structures, and also to simplify some of those structures. Food processors will be looking for competitively priced, improved-performance monolayer structures to replace some of their multi-layer plastic containers. These are challenges the industry must confront.

PET will continue to diversify and become the plastic of choice for many applications beyond soft drink bottles, because of all of the plastic containers in use today, the PET beverage bottle has the best collection and recycling network in place. Other PET containers can take advantage of that existing network.

New uses and markets for recycled plastic materials will be successfully cultivated. That's another crucial challenge. And, even if some varieties of plastic containers aren't ever able to be reincarnated over and over again into new versions of themselves in the same way and with the same efficiency that an aluminum can or glass bottle can be regenerated, that needn't be a negative.

Perhaps these plastic containers are recycled into a carpet. And once the carpet wears out, perhaps it is recycled into park benches. And once the park benches wear out, years and years later, perhaps they can be recycled into yet something else with an even greater landfill avoidance life.

During the decade of the 90s, the plastic packaging industry has the challenge and the opportunity to transform its _image_ of environmental archvillain into the _reality_ of environmental benefactor. I look forward to watching that happen, and I look forward to reporting on it. Thank you for your kind attention.

PLASTIC FOOD PACKAGING - A SOUND ENVIRONMENTAL CHOICE

David M. Adrion
Chief Environmental Officer
The Dial Corporation
111 West Clarendon Station 247
Phoenix, AZ 85077

Biography

"Dave" Adrion is Chief Environmental Officer for The Dial Corporation and is responsible for guiding Dial's efforts in the attainment of environmentally responsible products and operations.

Dave joined Dial in 1975 as a chemist in product development and has held positions of increasing responsibility in Dial's research and development, purchasing, and packaging development departments. His immediate past position was Director, Packaging Engineering.

Before joining Dial, Dave was with Proctor & Gamble Company in Cincinnati.

A native of Southwestern Ohio, he holds a B.S. degree in chemistry from Miami University (Ohio), and an M.B.A. from University of Phoenix.

Abstract

Dial helped pioneer the microwaveable meal product category with it's Lunch Bucket brand. This category represents a new and rapidly growing niche for plastics in food packaging. Now plastic packaging of all kinds is being attacked, unjustly, as an environmental liability. As a result, the growth and perhaps very existence of the microwave meal category is threatened.

This presentation makes the case for plastics as an environmentally responsible food packaging alternative, and closes with an appeal to each member of the audience to get the relevant facts out to the opinion makers in their locality.

<u>PLASTIC FOOD PACKAGING - A SOUND ENVIRONMENTAL CHOICE</u>

DAVID M. ADRION

CHIEF ENVIRONMENTAL OFFICER

THE DIAL CORPORATION

FOODPLAS '91 CONFERENCE

MARCH 5, 1991

ORLANDO, FLORIDA

INTRODUCTION

The Dial Corporation is proud of its pioneering role in the development of shelf stable microwaveable meals. In 1987, Dial's Lunch Bucket was the first major entree in the category. By 1990, just three years later, the shelf stable microwave meal category had grown to over $350 million in sales. Among the newer entrants is Dial's Light Balance line targeted at health-conscious consumers.

We believe this category has a very bright future. And it owes it's very existence to the multilayer high barrier plastic retortable microwave-transparent food container. Plastic was chosen as the packaging material of choice for this product category because plastic is clean, durable, protective, light weight, non-breakable and microwave-transparent.

Clean -	It imparts no flavor of it's own to the food product
Durable -	It holds up to high temperature, high pressure retort processing and the rigors of product distribution

Protective –	It is an excellent oxygen and moisture barrier
Light weight –	The plastic container is easy to carry to work, school, on trips
Nonbreakable –	It is safe for young children to handle, and conveniently portable
Microwave transparent –	It allows the product to heat quickly, and is totally safe for all microwave ovens.

In today's climate of heightened environmental sensitivity, plastics packaging is being unfairly singled out as a primary culprit in the solid waste disposal problem. We hear how plastic packaging is "wasteful of non-replenishable petrochemical resources". We hear how it is "not recyclable". We hear how it is "not degradable and lasts for hundreds of years, filling our landfills to overflowing". And, we hear that if its incinerated, "it fills the air with toxic pollutants".

Although these beliefs are flawed, the public's perception is that they are fact. And these "beliefs" held by the public to be "facts" have a strong influence on developing public policy. Left unchecked, the bright future of plastic food packaging will be compromised. This is an eventuality we must work to avoid through education and communication.

It is imperative that the plastics industry be effective in reaching national and local opinion makers with accurate information about the environmental properties of plastic packaging. It won't be easy. If it were, McDonalds would still be selling Big Mac's in polystyrene foam clam shells! But we have no choice. The alternative is to risk being legislated off the shelf as the result of unrelenting pressure from those who care deeply about the environment, but are uninformed about packaging.

I am sure that most of you are familiar with the EPA "Hierarchy of Integrated Solid Waste Management":

> Source Reduction
>
> Recycling
>
> Combustion with energy recovery
>
> Landfilling

By working through the hierarchy from top to bottom, an environmentally responsible waste minimization strategy can be devised to fit the peculiarities of any situation. So, if we can show that plastics are compatible with the EPA's integrated waste management strategy, it is possible to conclude that plastics are in fact "environmentally responsible" food packaging materials. Let's look at the facts.

SOURCE REDUCTION

Effective source reduction slows the depletion of environmental resources and prolongs the life of available waste management capacity. How does plastic packaging stack up as an aid to source reduction? Very well, I think.

Source reduction has been one of the major drivers in the long term growth of the use of plastics in packaging. When plastics replace other materials, total package weight often is reduced, as is, frequently, total package volume. And plastic packages have a way of getting lighter and lighter over time. For instance, the typical 2 liter PET soda bottle is not only much lighter than the glass bottle it replaces, but has itself been reduced in weight by 20% since it was first introduced in U.S. markets.

Now, I don't mean to imply that using plastics **always** results in net material savings, but that often **is the case.** Often enough that if the world stopped using plastic in packaging today, the total solid waste problem would undoubtedly increase. Therefore the net impact of plastic packaging on total solid waste volume is favorable. Plastic is a clearly an important aid to source reduction.

RECYCLING

Recycling is an extremely powerful tool in the arsenal of solid waste solutions. Recycling prevents potentially useful materials from being combusted or landfilled, thereby preserving waste disposal capacity and at the same time recapturing the materials for beneficial use.

Practically all plastic packaging is technically recyclable. But commercial success requires more than technical feasibility. It also requires a committed public, concentrated levels of materials, an infrastructure for recovery, well developed secondary material markets, and the economic "drive wheel" of reasonable material value.

Some plastic food packaging, notably polyester soda bottles and HDPE milk jugs, have already been proven to be good recycling candidates with ongoing and successful large scale recycling programs. And, Dial is participating in an industry program to prove the recyclability of rigid multilayer containers such as the Lunch Bucket bowl.

There is no question that industry is committed, as it should be, to driving recycling forward. It _is_ the elegant solution. But are any of you bothered, as I am, when you see these wonderful, light-weight, composite flexible structures, that have provided enormous source reduction benefits, being hammered because they are perceived as not

72

recyclable? It's a big concern. And there are other
examples of plastic food packages that don't fit the
criteria of good recycling candidates.

The plain truth is that recycling is not the only answer to
the solid waste problem, and sometimes it's not even the
best answer. There are times when recycling is neither
economically nor environmentally justified.

For its own good, industry needs to present a balanced view
to the public. There are four principal tools of solid
waste management, not just recycling. If we do not make
this clear, the public will insist on "recycling or else".

We surely need to support recycling. But when we talk with
our customers and to the public at large, we must remind
them that recycling is only one part of an integrated waste
management system. Recycling needs to be put into
perspective with the rest of the waste management hierarchy.

Before moving on, I'd like to cover one other recycling
issue. There are those who are apparently trying to
establish a hierarchy of recycling based on the notion that
it is somehow more "pure" to recycle over and over again
into the same container type than it is to use the recovered
materials in other durable or non-durable items.

This is a particularly difficult issue for the food industry, because there are proper and severe restrictions to the use of secondary materials in food contact packaging.

Frankly, I think the idea of a recycling hierarchy is nonsense! The point is to find high value uses for the recycled material. Does it really matter whether the material goes back into containers or is put to some other useful purpose? Let's see what we can do to nip this misguided idea in the bud.

COMBUSTION WITH ENERGY RECOVERY

While recycling has received more than its fair share of public attention as a solid waste management strategy, combustion with energy recovery, also called waste-to-energy incineration, has been virtually ignored.

Clearly, some people are vocally and adamantly opposed to incinerators. But this appears to be a particularly U. S. based phenomena. In the rest of the developed world, incinerators are far more common. Those countries which are most successfully managing their solid waste problem rely heavily on waste-to-energy incineration.

It's not exactly fair to say that industry has completely ignored the combustion option in its solid waste management dialogue, but industry has shown little inclination to promote the option. It seems as if, in our conscientious effort to respond responsibly to the recycling option, the combustion option was lost in the shuffle.

I also think we've been reluctant to engage in what could be acrimonious dialogue with those who are strongly opposed to incineration. After all, we'd like to have some of them as our good customers.

Promoting incineration could be difficult. An article in the most recent quarterly publication from GREENPEACE enumerates ten strategies for opposing incinerators. Among other things, GREENPEACE activists were directed to be firm right from the start in their objective to prevent construction of incinerators. Not to negotiate. Not to dialogue. Not to reason. But to prevent!

It is clear that there are some people who will blindly oppose incinerators whenever and wherever proposed. But we still need to patiently and repeatedly bring the facts to light. In Japan, 34% of all solid waste is incinerated. Waste-to-energy incineration is also common in many European countries. This is not because these countries are lacking in environmental sensitivity, but rather because modern, properly operated incinerators work. They do not "poison the environment". They do generate needed power. And, solid waste cannot be managed effectively without them.

That portion of plastic food packaging waste that isn't easily recycled is an ideal candidate for waste-to-energy incineration. Plastics from food packaging burn cleanly and have an extremely high energy value. This is because when plastics are manufactured from natural gas and petroleum, the energy in the starting material is preserved.

In Germany, when they talk about waste-to-energy incineration they refer to it as "harvesting" the energy content of the waste. In the case of plastics, I like to think of it as "returning" the energy that was "borrowed" when the natural gas and/or petroleum was taken to produce the plastic.

After all, the vast majority of the natural gas and petroleum taken from the ground goes directly into energy production. We take it from the ground and we burn it. So what's wrong with diverting a small portion, using it to make a bottle, and then, when we are finished using the bottle, burning it to recapture the energy. It seems to me this only adds to the usefulness of the extracted resources.

I was surprised to see in last summer's EPA figures a fairly substantial increase in the amount of waste-to-energy incineration in the United States between 1986 and 1988.

Clearly, new plants are being approved and built. We in the plastics industry and those of us who use plastic packaging need to support this trend.

LANDFILLS

Where it is not possible to recycle or harvest the energy content, landfilling is the disposal option of last resort. Even though landfilling is last on the solid waste management hierarchy, it is the predominant disposal method in use in the United States today.

When plastic food packaging is landfilled, it is stable and completely non-toxic. It contributes neither to subsidence nor to gas or leachate formation. The inert nature of plastic makes it an ideal candidate for disposal by landfill.

CONCLUSION

Plastic is an environmentally responsible choice for food packaging. It is a good material for achieving source reduction, it is recyclable, it has high energy value when combusted for energy ,and it does not produce toxic byproducts in a landfill. These are the simple facts, we need to get across to public opinion makers (as well as to the press). In closing, I ask each of you to get involved in spreading the word. You will make a difference!

MICROWAVE PACKAGING AT GOLDEN VALLEY

Sara J. Risch, Ph.D.
Director Research and Development
Golden Valley Microwave Foods
6866 Washington Ave. S.
Eden Prairie, MN 55344

Biography

Sara Risch is currently Director of Research and Development for Golden Valley Microwave Foods, Inc. Before joining Golden Valley in 1988, she was a research associate in the Department of Food and Nutrition at the University of Minnesota and a principal in the consulting firm of R & R Analytical, Inc. Sara worked with food, flavor and packaging companies on issues including encapsulation of flavors, microwave flavors and food-package interactions. She received a Bachelor of Science degree from the University of Minnesota, a Master of Science from the University of Georgia and a Ph.D. from the University of Minnesota. All the degrees were in Food Science. Sara frequently speaks at conferences and symposia, has published numerous articles and is editor of a book entitled Flavor Encapsulation.

Abstract

Golden Valley Microwave Foods, Inc. is committed to developing microwave only food products. An integral part of microwave foods is packaging which will allow proper cooking. Golden Valley uses predominantly flexible materials comprised of paper and metallized film. The type of paper used is usually a combination of bleached Kraft and greaseproof which is critical for processing, performance and product safety. The film that is metallized is typically polyester, however, other types of films have been investigated. Another key component of the products is the overwrap. This film needs to provide both moisture and aroma barriers. While the same type of film can be used for all products, different properties are desirable for frozen versus shelf stable foods.

PACKAGING FOR
GOLDEN VALLEY MICROWAVE FOODS

FLEXIBLE PACKAGING

- Paper
- Film
- Laminates
- Inks
- Adhesives

80

LAMINATES

- Popcorn
- French Fries
- Waffles
- Other

FILMS

- **Machinability**
- **Seal Quality**
  - Hot Seal
  - Cold Seal
- **Stress Resistance**

FILMS FOR OVERWRAP

- **Aroma Barrier**
  - Good Aroma In
  - Off Aromas Out
- **Moisture Barrier**
- **Temperature Resistance**

FILMS FOR SUSCEPTORS

- **Heat Stable**
- **Barrier Properties**
  - Improve Shelf-life
  - Eliminate Overwrap
- **Cost**

PAPER

- Greaseproof
- Strength
- Barrier Properties
- Release Properties

INKS

- **Microwave Packages**
  - Water Soluable
  - Low Residual Volatiles
  - High Print Quality
- **Overwraps**
  - Scuff resistance
  - High Print Quality
  - Low Odor

ADHESIVES

- Water Based
- Low Odor
- Heat Seal
- Cold Seal

FRENCH FRIES

- **Overwrap**
  - Printed
  - Unprinted
- **Carton**
- **Condiments**
  - Salt
  - Ketchup

FRENCH FRIES - SUSCEPTORS

- Large Surface Area
- High Moisture
- Good Adhesive
- Uniform Heating

SOLID WASTE ISSUES

- Volumes
- Recyclability
- Biodegradability
- Use Of Recycled Materials

CHALLENGES

- Cost Reduction
- Improved Performance
- Safety
- New Applications

FOOD PACKAGING: IMPLICATIONS AND CONSIDERATIONS
FOR RECYCLED PLASTICS

M. Jaye Nagle
Director, Scientific Relations
Kraft General Foods
801 Waukegan Rd.
Glenview, IL 60025

Biography

Ms. M. Jaye Nagle is currently Director, Scientific Relations for Kraft General Foods. In this capacity, Jaye is responsible for proactively working within the KGF technical community and externally with key regulatory, academic, trade and consumer groups to interface on topical scientific issues of significance to KGF. Specifically, Jaye has Scientific Relations linkage responsibility for the three Glenview-based business units: Kraft USA, KGF Frozen Products and KGF Commercial Products. She will have lead responsibility for scientific issues related to Solid Waste.

Jaye received her B.S. degree in Biological Science from Northern Illinois University in 1975 and her M.S. degree in Food Science from the University of Illinois Champaign, in 1977. Her M.S. research was in Food Microbiology, specifically characterizing cell injury including subcellular mechanisms of Staphylococcus aureus following exposure to sublethal acid environments.

Jaye began her career with General Foods as an Associate Food Technologist in 1977 in the Beverage and Breakfast Food Division working out of the Chicago Kool-Aid facility. While at GF, she gained Product Development, Technical services and Packaging experience on powdered beverages.

Jaye joined Kraft in 1986 as a Senior Research Scientist II after spending 4 years at Quaker Oats Technical Center in a variety of Packaging Development project and management roles. At Kraft, she progressed through a range of product and package development assignments and in her most recent position was Associate Director for the Grocery Products Division of Kraft USA.

Jaye currently chairs the KGF Solid Waste Task Force, and is working in several trade committees and task forces to collaborate industry efforts against Solid Waste initiatives.

Abstract

Recycling, in general, is receiving increased attention as environmental considerations continue to play an ever-increasing role in all forms of packaging, including food packaging. This presentation will discuss some of the considerations for the potential uses of recycled plastics in food packaging. Special emphasis will be placed on the safety, purity and quality considerations necessarily involved in food packaging. Additionally, some background information highlighting the functionality and special role that food packaging plays in assuring a safe and wholesome food supply will be discussed.

Food Packaging: Implications and Considerations for Recycled Plastics

for presentation at
FOODPLAS '91

M. Jaye Nagle
Kraft General Foods
March 5, 1991

The environment has become one of the world's most pressing
concerns, leading many to call the 1990's the Decade of the
Environment. The issues we face are many: depletion of the
ozone layer, acid rain, the greenhouse effect, air, water
and land quality, deforestation, depletion of natural
resources and last, but definitely not least, solid waste.
The solid waste issue can best be summarized by two simple
facts: First, Americans are generating more and more trash.
The United States Environmental Protection Agency has
estimated that we are currently generating over 180 million
tons of municipal solid waste per year -- that's 4.5 pounds
for each man, woman and child every day!! The second fact
is that landfill space, where most of our solid waste
currently ends up, is decreasing. Almost half of the now
available landfills are expected to close by the end of
1991. And, it's becoming increasingly difficult to site new
landfills due to strict regulations and strong municipal and
public pressures against them.

Food packaging, in particular plastic packaging, is a very
visible component of our municipal solid waste stream and is
receiving an increasing amount of scrutiny. Food and
beverage packaging accounts for approximately two-thirds of
the $70 billion packaging industry in the U.S. When
consumers ask themselves what they can do to help solve the
solid waste crisis, packaging is the element over which they

95

feel they have the most control and can personally impact solid waste. It's not surprising, when you consider that every day, at every meal at home, and in many meals away from home, consumers see the packaging used to contain their food get thrown away.

It is ironic that it is the very packaging that is often criticized for environmental reasons that guarantees the safety along with product integrity and quality of the many products that we enjoy. Modern packaging and plastic packaging, in particular, is absolutely essential to delivering our mass food supply and essential to the health, welfare and convenience of American consumers and families. However, most consumers may not appreciate all of the functions and benefits packaging has provided in delivering their food to them safely every day. For this reason, many consumers are demanding that all packages, including food packages be designed to have less impact on the environment.

Individual companies and industry and trade groups are responding to this need and are continuing to demonstrate a sense of urgency in being actively involved as part of the solution. Within the plastic packaging industry, groups such as the Council for Solid Waste Solutions (CSWS), the Council for Plastics and Packaging in the Environment (COPPE), the Plastics Recycling Foundation (PRF), the Society of Plastics Institute, and others have supported

research, demonstration projects and other positive actions
which have helped to increase awareness and begin to develop
a framework for industry collaboration.

I would like to take this opportunity to discuss an emerging
environmental topic -- the future opportunities for
plastics recycling into food packages. The concept of
recycling plastics back into food packages is one of the
"hottest" topics around today. It is truly exciting for
both the plastic and the food industries and is receiving
significant media attention. Before I get into that,
however, I think it's important to first review the unique
role of food packaging, specifically highlighting the
important part food packaging plays in reducing food waste.

Food is perhaps the most perishable item which is purchased
by consumers. It is subject to spoilage from the moment of
harvest, slaughter and manufacture. It is also subject to
infestation and attack by microbes, rodents and other pests.
In today's society, the use of protective packaging allows
food to be shipped over great distances and to stay fresh
and unspoiled for long periods of time.

In the United States today, food spoilage is less than 3%
for processed food and 10-15% for fresh food. In lesser
developed countries, food spoilage due to inadequate or
non-existent packaging can reach 50%. In a recent study by

the United Nations International Trade Center, results
showed that some 30% of the export earnings of developing
countries are lost due to inadequate packaging, resulting in
severe product losses. Similarly, a recent study by Dr.
William Rathje, a University of Arizona archaeologist found
that Mexico City residents produce about 40 percent more
garbage per household than do their US counterparts. One of
the reasons appears to be the use of considerably less
packaging and prepared foods in Mexico. Mexican waste
contains a great deal more organic material and more spoiled
food. In typical US food processing operations many of the
by-products -- items such as peels, hulls, leaves,
trimmings, etc. can be converted into useful products such
as animal feed, or can be returned to the ground as humus.

Another example illustrating this concept often cited is
orange juice. A typical 64 ounce package of ready-to-drink
orange juice contains the juice of 20 oranges. In
manufacturing the juice, the food processor recovers the
peels for use as animal feed. If the consumer were to make
the juice at home, he or she would have discarded the peels
weighing approximately 35 ounces or a little over two
pounds. In this one example, there is a solid waste
avoidance of approximately 90% in use of the packaged orange
juice, not to mention the economic gain of converting the
peels into a valuable by-product. If one were to

extrapolate this example to the city of New York, with a population of approximately 8 million, and assume half the population were to have one 6 ounce glass of orange juice a day, there would be a net increase of 900,000 pounds of solid waste per day just from the peels that would be discarded at home thereby entering the municipal solid waste stream.

Thus, we can see that while packaging may end up as waste once it has done its job, the fact is that in doing its job, packaging often prevents infinitely more waste than it creates. Here is a dramatic graphic representation of this relationship between increased use of plastic packaging and a decrease in the percent of food discarded in the municipal solid waste stream.

Little public attention is paid to post harvest losses at the various stages of the distribution chain. More simply put, the many benefits of modern packaging are often overlooked and taken for granted. Take away the packaging, take away the protection it affords, and the inevitable result will be a huge increase in microbial and physical damage... damage that constitutes food waste.

Let's take a quick look at the many functions of a package to better understand how it is that packaging plays this valuable role in our everyday lives.

First, and foremost, we need packaging to contain products.
Any commodity which is not contained is extremely difficult,
if not impossible, to transport. Packaging is an
indispensable element in the physical distribution of goods,
the system by which agricultural and manufactured foods are
moved from the farm and the factory to the consumer. In a
very real sense, packaging has completed a big portion of
its job by the time the food product is in the consumer's
hands.

Even in prehistoric times, nomadic tribes had to rely on
packaging ...animal skins, gourds, containers made of cane
and bark ... to contain their meager supplies of food and
water when they moved to a new location. Less crude forms
of packaging evolved: woven baskets, earthenware pots from
river clay, and with the advent of the Industrial Revolution
our society changed from a rural and agricultural society to
an urban and industrialized one. The food distribution
chain lengthened as the point of food production became
geographically separated from the point of consumption.
Only 3% of Americans live and work on farms today. The fact
that these few Americans can feed a nation of 250 million
people and millions more overseas is due in part to the
success of modern distribution systems and packaging. The
whole system of mechanization depends upon packaging to
contain the products being transported and distributed.

Food packaging's role goes beyond simple containment, however. It must also protect the food product so that it reaches the consumer in an acceptable condition, as close as possible to the condition when it left the farm or factory. All foods are subject to spoilage of some type. As food packagers know, knowledge of the barrier properties of a package material are essential to development of an economical and effective protective package design for a given product. Chemical barriers can provide protection from light (visible and ultraviolet), oxygen, moisture, aroma (both maintaining desired aromas as well as preventing off odors), carbon dioxide and other gases. Besides providing protection from chemical barriers, the package also provides protection from microbiological and physical challenges, including infestation by insects and rodents. In particular, food packaging which comes in direct contact with the food must be wholesome, safe and pure, not allowing any off-odor or flavors to be transmitted to the food product during the entire shelf life, including through final end-use by the consumer. Normally, no one packaging material alone can satisfy all of the protection and preservation requirements for even one individual food product. Add to that the plethora of food products in the market today and the explosion of new products and line extensions, and it is easy to see that the selection of the correct packaging material requires the combined skill of knowledgeable food and packaging engineers working closely together.

In addition to containment and protection, product safety, is of paramount importance to consumers. A very critical aspect of packaging's protective role is that it makes a product safe for consumers to use. By protecting foods from microbiological attack and spoilage, packaging has virtually eliminated the risk of outbreaks of food-borne disease that confronted earlier generations. While packaging protects food from spoilage and contamination, it also protects the food product from tampering. Clearly within the last decade many food packages have added tamper-evident features to help provide visual assurance to the consumer of package integrity.

Packaging also plays a large role in communicating about the product. A typical American supermarket contains tens of thousands of food and consumer items. In addition to attracting the attention of the shoppers and providing brand identification, packaging conveys vital information of interest to consumers, such as ingredient labeling, nutrition and health information and other information about the food attributes. This information help consumers distinguish among products and aid in purchase decisions.

Finally, a growing number of packages are satisfying consumer demands for a range of choices, which include value-added options for functionality and time saving convenience options. Desired features include: microwaveable, table-ready, easy-open, squeezeable, recloseable, reuseable and single-use packages are increasingly sought by finicky consumers. Modern packaging, drawing heavily upon plastic packaging technology provides unequaled options and choices available to US consumers today, and, in so doing, has literally transformed consumer lifestyles. Among the many benefits of plastic packaging are: Lightweight, ease of handling, durability, strength, shatter-resistance, design versatility, and relatively low cost.

It bears repeating that modern packaging and plastic packaging, in particular, are essential in delivering our mass food supply and are essential to the health, welfare and convenience of the American consumer today.

Let's move now to a discussion of some of the environmental issues facing packaging today. Not only is the package the "salesman" for the product, it is increasingly becoming the "environmental conscience" as well. Consumers are increasingly demanding more information about the environmental attributes of packages and are seeking "green" in all that they do. Though they want the package label to include information about the "greenness" of the package, the vast majority of consumers don't know what it is that makes a package environmentally sound and often make judgments about packaging based upon inadequate and misinformation, quite often -- and simply put -- misperceptions. Responsible food manufacturers are working together to provide national, uniform guidelines around the environmental labeling of products and packages. The National Food Processors Association (NFPA) will be petitioning the Federal Trade Commission in the first quarter of 1991 with a document supporting the adoption of responsible environmental labeling guidelines as the basis for a self-regulated industry process.

Plastic packaging, in particular, has been the target of much debate and has been singled out by many as the major cause of the municipal solid waste problem currently facing the US. That perception, and the growing sense of urgency and call to action within all facets of the plastic packaging industry have caused plastic recycling to emerge recently as a focus area of interest and activity.

The ability to use recycled plastics in food containers represents very real and pressing environmental benefits and perhaps long term economic incentives as well. Let's focus now on the opportunities associated with plastics recycling for food packages.

Today, based upon the latest Information Included In the EPA Agenda for Action, 1990 Update, recycling accounts for about 13% of solid waste management in the US. Of that 13%, 7.5%, or 58% of the total amount of material recycled is all forms of packaging. Currently, most of the recycled packaging materials are glass and metals, both aluminum and steel. The amount of plastic packaging material which is currently recycled is low, in the order of 1-2%. The two major packaging resins which are being recycled to an appreciable degree from post-consumer waste In the US today are polyethylene terephthalate (PET) and high density polyethylene (HDPE). These have been mainly in the form of plastic soft drink bottles and milk jugs. It is estimated that there are between 1500-2000 curbside recycling programs collecting post-consumer waste in the US, with approximately 400 of them Including some plastics. Of the plastic containers that were collected in 1990 It is estimated that about 300 millions pounds were PET and 200 million pounds were HDPE.

Compared to many other well established recycling systems, plastics recycling is in its infancy. It has been said that we will know when plastics recycling has finally arrived when the amount of plastics being recycled are measured in tons, not pounds.

A number of new projects, however, do hold promise for greatly expanding plastics recycling in the future. For example, the National Polystyrene Recycling Company was formed in 1989 and has set a goal of recycling 250 million pounds, or 25% of post-consumer polystyrene products by 1995. This is in keeping with the EPA goal of 25% waste reduction. In order to achieve this, large regional polystyrene recycling plants are being built across the country, and an infrastructure to recycle foamed polystyrene packaging from schools, restaurants and other institutions is being developed.

In order for significant development and hope of attaining national recycling rates suggested by EPA for all plastics, collaborative efforts must be coordinated with the active support and involvement of communities, corporations and individual citizens. An example of just such an effort was the Minneapolis pilot plastics recycling program which began in 1990 following proposed bans on selected packaging materials, including rigid plastic containers. The NFPA and the Council for Solid Waste Solutions (CSWS) worked together with the appropriate community organizations to help establish systems to include plastics in the curbside recycling program. Results were very successful and the viability of the program was demonstrated. This is the type of success story which needs to be expanded and built upon as the plastics recycling industry develops. And while I, personally, believe the future of plastics recycling is very bright, there is, of course, no one answer, no "silver bullet" which will solve the complex issues we face associated with municipal solid waste.

What are the recycled plastics being used for?

The recycled materials have typically been used in a variety of uses including carpet fibers, fiberfill, high performance engineered plastics, park benches, plastic lumber and more. Within the last year, consumer products companies have found ways to incorporate a layer of recycled HDPE plastic into their multilayer detergent and household products bottles, in levels from 25%-35%. There are also containers made of 100% recycled PET, for example, the Spic N Span bottle. Most recently, Proctor & Gamble, has also just introduced recycled HDPE into a layer of the flexible overwrap used to package their disposable diapers.

The first food application was last year in an egg carton made with recycled polystyrene, admittedly a very unique direct food contact application since the eggshell provides a natural barrier of its own.

In thinking about the potential markets and different uses for recycled plastics in food packages, let's look at the wide variety of foods now packaged in rigid plastics.

PET...

HDPE...

LDPE...

PP...

PVC...

PS...

Multilayer...

By being able to use recycled materials in food packages,
new higher value markets will open up to recycled plastics,
stimulating recycling and decreasing the amount of virgin
raw materials used, thus decreasing the waste that society
sends to landfills and incinerators.

To recognize this opportunity, however, a significant
challenge awaits us -- that of developing technology that
will guarantee the use of recycled plastics will retain the
food product in a safe and wholesome manner. This means
that the risk of any food safety issues associated with
possible contamination through the post-consumer collection
process need to be negligible for any materials which are
used in direct food contact applications.

Potential contaminants could result from three main areas:

1. Contents residuals from food packaging. This could be
from a beverage, as in the case of milk and soft drinks, or
it could be from any variety of foods which could ultimately

end up in the plastic recycle stream when collection systems
go beyond beverage bottles.

2. Contamination from non-food packaging. This could
include any of a variety of household products which may be
packaged in plastic bottles and ultimately included in the
plastic recycle stream.

3. Incidental contamination from the use of plastic food
containers for storage of other products.

Let's look at each one of these potential sources of
contamination with a bit more detail. First, contents
residue from food packaging. To appreciate this, it is
important for us to look at the wide variety of types of
food products packaged in plastics and the varying
characteristics they can have.

Dairy products

Liquids (soft drinks, juices)

Bakery goods

Dry grocery items

Meats, prepared foods

Cooking oils and spreads

These various products represent very complex systems, in
some cases biological products whose composition will
fluctuate seasonally. These different food systems will

have different levels of acidity, fat and oil content, moisture, and other soluble materials. It is unknown if the currently employed physical and mechanical cleaning and processing systems used by the plastics recycling operations will be able to clean these products effectively and easily.

The second area of potential contamination, that from non-food packaging, could contain any variety of household products such as bleach, detergents, cleansers, motor oil, and so on. In addition, contamination from the colorants and stabilizers used which are not approved for direct food contact could come in contact with food containers. Again, it is uncertain what level of contamination will remain in the post-consumer stream and what effect this would have on the potential use for food packages.

The third area of potential contamination, from incidental contamination could result from intermediate use of the food container for storage of materials other than food, or from the co-mingling of the empty food packages within the recycling bin or within the collection and handling system with other potentially toxic substances. Possible examples could include materials such as decaying food, gasoline, kerosene, pesticides, household chemicals of many types as well as contamination from common insects and rodents. This incidental contamination, overall, is probably the biggest area of concern and the one which would need to be carefully evaluated in future investigations.

In addition to maintaining chemical integrity and purity, in order for the plastic material to be able to be recycled, the physical properties and performance of the recycled material through its ultimate end use by the consumer must be proven. And, we must not forget that these plastic recycled materials will need to continue to be functional as their recycled life extends over and over again. In other words, they need to be re-recyclable, and thus, need to be able to withstand multiple heating histories without impeding barrier or functional properties or any other important performance characteristic. Acceptable sensory performance would also be included among this list of performance attributes. Certainly for a food product this becomes paramount -- assuring that the recycled materials do not impart off-odors or off-flavors to the food product.

Additionally, for any material to be significantly used for recycling of food packages, it must be available in the recycle stream in sufficient quality and quantity through existing infrastructures to meet the required demand. With the current state of plastic recycling, this most likely would currently apply to PET and HDPE. Of course, this could and is likely to change in the future, as plastic recycling systems develop and consumer education and

participation grows commensurately. Other plastics have
been shown in pilot and demonstration projects, from a
technological perspective, to be capable of being recycled.
However, it is the ultimate market demand driven by
economics which will determine the future of recycled
plastics.

Earlier I mentioned the recent press announcements by
Coca-Cola and Pepsi, in conjunction with Hoechst Celanese
and Goodyear, respectively, which reference the use of
various depolymerization and repolymerization processes.
Similarly, other major PET resin suppliers, Eastman and
Dupont have also announced through separate press releases
some details of their approaches to depolymerization. These
announcements stimulated great interest and enthusiasm
around plastic recycling and have readied the entire
industry for action.

Certainly the relative expediency with which the FDA has
responded through a letter of no objection to the Hoechst
Celanese process is encouraging. The FDA, in their letter
of no objection, specifically state that they view the
Hoechst process as "regeneration", which one would interpret
as a very specialized case of recycling. This is a key
distinction between this chemical approach and other more
conventional, but as yet unproven physical and mechanical
processes.

Although each resin company's approach so far is different
and each company is reluctant to divulge process details,
they all share a common "unzipping" approach by chemical
unraveling and rebuilding of the polymer. Some of these
processes reportedly go back further in the process,
however, they all have a chemical basis for purification
which could include filtration, distillation, and/or
crystallization and other mechanisms. These recycled or
regenerated materials are then reacted back into polymers
that reportedly have equivalent properties and purity to
those made from conventional virgin raw materials.

At least one key unanswered question, however, remains the
long term economics of these systems. Currently, capacity
is limited and reportedly far short of market demand. Costs
do reflect this supply/demand situation, hence, pricing for
the regenerated polymers are in the range of 10-15% upcharge
over virgin resins. It will be interesting to see if this
trend continues in the long term and ultimately, what price
the consumer packaging market will bear.

Lets consider another possibility for future plastics
recycling which does not use one of the depolymerization
processes. One possibility would be to exclusively use
physical and mechanical processes to purify through
cleaning, filtration and density separation techniques, that
is, using the systems currently in place in plastic

recycling processes. While this many sound a little unlikely, and may raise questions of safety and purity, this would essentially use the approach currently used by the recycled paper Industry. With sound scientific protocols, comprehensive data and potentially new techniques, it is not out of the realm of possibility that such a system could be devised and proved to of a purity suitable for direct food contact. There have been discussions and proposals within the Industry around this, and it is likely that some level of research will be conducted in this area. Certainly the economics of such a system, devoid of complex chemical reactions and reactors would have a benefit to the de- and re-polymerization approaches. I believe this will be an Important area in the future for the food and consumer products Industry.

In summary, I hope I have conveyed that food packaging plays a vital role in the safe production, distribution and use of our abundant American food supply today. The multitude of food packages used today ensure a variety of food choices which are safe, wholesome and of the highest quality. Food manufacturers like Kraft General Foods are meeting the challenges of solid waste management through proactive support of environmental Initiatives and through package design processes which support an integrated approach to municipal solid waste, based upon the EPA hierarchy of

source reduction, recycling, composting, waste-to-energy
incineration and safe landfilling.

The food industry is faced with developing packages that fit
modern American lifestyles, packages that offer convenience,
value, quality, ease of use, protection of products and
which are environmentally sound. Food packagers are
committed to balancing these economic and social
considerations with environmental issues to produce packages
which meet not only the use, but also the disposal needs of
society.

BANQUET SPEAKER

HARRY E. TEASLEY

ON THINKING ABOUT PACKAGING AND OUR ENVIRONMENT

Harry E. Teasley, Jr.
President and Chief Executive Officer
Coca-Cola Foods, a Division of
The Coca-Cola Company
Vice President
The Coca-Cola Company
P.O. Box 2079
Houston, TX 77252

Biography

Harry E. Teasley, Jr., is president and chief executive officer of Coca-Cola Foods, a Division of The Coca-Cola Company, and a vice president of The Coca-Cola Company. Coca-Cola Foods, headquartered in Houston, Texas, is the nation's leading producer and marketer of fruit juices and drinks. Its products include Minute Maid brand fruit juices, punches and ades, Five Alive brand citrus beverages, Hi-C brand fruit drinks and Bacardi tropical fruit mixers.

In addition, Mr. Teasley also serves as Chairman of The Coca-Cola Company's corporate worldwide task force dealing with environmental issues. He is active on a number of industry and inter-industry committees dealing with public policy and the environment.

Before assuming his current position with Coca-Cola Foods in April 1987, Mr. Teasley was managing director of the Company's soft drink bottling operations in southern England.

A graduate of the Georgia Institute of Technology with a degree in Industrial Engineering, Mr. Teasley joined the Coca-Cola Company in 1961 as a senior engineer with the Technical Research and Development Department. At various times in his career, he has held responsibilities in the areas of packaging, new products, sales equipment, distribution systems, management science and corporate business development in the merger, acquisition and venture areas.

THINKING ABOUT PACKAGING,

THE ENVIRONMENT,

AND PUBLIC POLICY

Harry E. Teasley, Jr.

President & CEO

Coca-Cola Foods

Beginning in the late 1960's, packaging and other products began to come under fire for their contribution to litter, solid waste, and resource depletion; and that pressure has certainly intensified over the last couple of years.   The pressure has focused primarily, not totally, but primarily, on solid waste.

Since solid waste is one of the major driving forces being used to justify various forms of packaging regulation, it is appropriate to begin our discussion with a brief examination of the solid waste issue.

Many of the articles that we see and much of the public rhetoric describe a solid waste crisis.   If there is a crisis, what is the nature of that crisis?   Is it that we are producing too many products in this society and we should scale back our consumption and standard of living?   I think not.

Do we over-package and use materials in a profligate manner?   Again, I think not, and will so argue later in my presentation.

We read that the United States is running out of landfills, and that by the mid-1990's, there will be no place left to place the residual materials of human activity.   Can that be true?   Is that a correct and accurate statement of the situation?   Again, I think not.

At any point in time, the existing landfills have a rather finite life, or at least finite capacity.   That's what one should expect.   That's always been the condition, and so it is fair to say that at any time the existing set of landfills have a finite life or reserve capacity.

The traditional response has been to site new landfills. And it is here where the problem truly exists. I think the crisis is far more political than physical in nature. We live in a time where it is increasingly difficult to site and permit new landfills and/or incinerators.

It is not the availability of appropriate places to put residual materials ~ it is, rather, the political will to make those places available.

Enough on the driving force. That driving force has in the last few years promulgated a plethora of laws, regulations, and proposed laws and regulations, at the federal, state, and local levels. These laws are often contradictory, discriminatory, and ~ as I go through this presentation I will argue ~ often unlikely to meet their professed objectives.

Increasingly, the laws and regulations are directed at reducing the residual material in the economy from industrial and consumer activities. And increasingly, the laws and regulations are moving toward specifying materials and/or packaging, or banning materials and/or packages.

In addition to material and/or package specifications, bans, or mandatory refunds, there is also a major thrust to mandate recycling through legislation. I find the recycling issue truly interesting.

Historically, the thrust for recycling did not come from government or solid waste officials, or even environmentalists ~ it came from commerce and industry as business sought to become more efficient and reduce costs. Let me give you an example from my own company and my own industry ~ orange juice.

At Coca-Cola Foods we produce a number of orange juice products, from ready-to-serve to concentrated. We squeeze oranges for those products in Florida. In the process of squeezing those oranges, we generate a residual material ~ orange peels.

To dispose of that huge quantity of orange peel would be incredibly costly. Because we have a large quantity of a sorted material at one location, one of the people in the orange industry long before I was ever involved, figured out that orange peel could be efficiently and economically converted to animal feed.

So today, a typical orange processing plant not only produces animal feed but also recovers orange oil and d'limonene in the process as well, which are used as by-products for other processes.

Several years ago I was in a meeting where an individual argued that my company should not be producing pre-packaged orange juice. That individual thought that the orange was the greatest example of a product which required no packaging, and that was the way it should be sold.

But let's examine the thesis being expressed by that individual. In the view of the individual, fresh oranges would appear to be an excellent example of a product that exemplifies the CONEG concept of <u>no packaging</u>.

Upon closer examination, this is not the case. Fresh oranges are packaged in a rather substantial telescoping corrugated container for distribution to retail outlets. But that is not the place to begin one's analysis.

Juice processors squeeze oranges more efficiently than consumers because of the industrial equipment which they use to perform the task. A consumer will, at a minimum, require at least 25 percent more oranges than a processor to yield the same amount of juice.

So, home squeezing of fresh oranges is less efficient of oranges and, therefore, less efficient of agricultural land, fertilizers, pesticides, water resources, agricultural capital, and agricultural labor than packaged orange juice.

Fresh oranges generate almost 9 times more corrugated waste at retail than does the 12-oz. frozen concentrate alternative. And, at the consumer level, fresh oranges generate over 60 times the poundage of waste as the 12-oz. frozen concentrate alternative. The consumer waste is, of course, wet peels versus the small composite can.

When a consumer squeezes oranges, the wet peels are disposed of through the solid waste collection and disposal system (except in the situation where the consumers compost their kitchen wastes) while, as already noted, a juice processor converts the peel to animal feed.

The fresh orange alternative also weighs about 7.5 times as much as the 12-oz. frozen concentrated alternative, and requires about 6.5 times as many trucks to distribute equal quantities of orange juice to the consumer. So, in addition to agricultural efficiencies, the 12-oz. frozen concentrated orange juice produced is more efficient of trucks, diesel fuel, and road systems.

Well, the bottom line to that little story is that squeezing oranges industrially, and packaging either concentrate or ready-to-serve juice, is very efficient from a waste-produced standpoint, very efficient economically, and very efficient environmentally.

I am sure that some of you in this room could present similar examples from your own industry. You should know those examples ~ you should document them ~ you should understand your processes and your packages.

Now, let me get back to recycling, but in a more general vein.

In his essay, _First and Second Things_, C. S. Lewis ~ classicist, Cambridge don and essayist ~ advances the thesis that people, in striving for second things rather than something more fundamental ~ first things, achieve neither the second nor the first thing.

In his essay with the following quote ...

> _"I think many would now agree that a foreign policy dominated by a desire for peace, is one of the many roads that lead to war."_

... Lewis makes the point that peace is not a first thing and those who strive as if it were, will probably be disappointed.

Is recycling a first or second thing?

I argue that recycling is a second thing and a tactic or process, to achieve a more fundamental goal or objective.   A more fundamental goal or objective might be the efficient use of some resource like:

- ◄ Materials
- ◄ Energy
- ◄ Landfill space, and
- ◄ Financial resources.

Focusing on recycling rather than the more fundamental reason for recycling may, in fact, cause us to fail in our quest to achieve the more fundamental goal.

Recycling is not free.   The process of recycling may require:

- ◄   The expenditure of energy
- ◄   The use of process materials
- ◄   Investment in capital facilities
- ◄   Labor inputs, and
- ◄   Financial resources.

A commitment to recycling must not be based on a religious view that recycling is per se virtuous and, therefore, desirable.   Rather, it seems, there must be some objective criteria for determining when it is appropriate and not appropriate to allocate resources to a recycling activity.

I argue that the best objective criteria is efficiency.

But that raises the question: Efficiency of what?

- ◄ Material?
- ◄ Labor?
- ◄ Capital?
- ◄ Landfill space?

One could make an argument that all are important and that we should strive for efficiency in all areas.

But we know from Bentham's law that we can maximize only one variable at a time. What variable do we maximize?

In a market economy we have established a single mechanism through which we attempt to make incommensurables commensurable, and to provide us with a mechanism for making these tradeoff decisions. That variable is, of course, economic cost.

Recycling is not new.   Rag merchants, scrap metal dealers, waste paper dealers and junk yard operators have long participated in recycling activities which were, in grand terms "economic" and in coarser terms, "where they could make a buck."

Many industrial operations and retail operations recycle effectively and efficiently because the economics are right.   The economics are often right because several conditions exist.

There is often a large amount of uniform, sorted material being generated at a point source.   These conditions of sorting, aggregation and single location, often give rise to economic recycling.

As products are dispersed from manufacturer through distributor, through retailer, and finally to consumer, these conditions are degraded. Products are disaggregated, mixed and dispersed. Recycling then, is dependent upon collection, aggregation, appropriate sorting, and transportation.

If an economic analysis indicates that it doesn't make financial sense to recycle in a given circumstance, then that should be a <u>red flag</u> because it means that the recycling process is consuming some set of costly resources ~ maybe energy ~ maybe labor ~ maybe capital ~ maybe other materials.

From a technical standpoint it is possible to recycle any material. At the grandest level we know that matter is neither created nor destroyed except for some matter/energy transformations in the nuclear domain.

This gives rise to the question of when do we recycle, or when is it appropriate to recycle?

This question brings me back to my earlier point, and the answer must be, when it is efficient to do so. And <u>efficient</u> probably has to be judged on the basis of economic efficiency because that is the only mechanism we have to render incommensurables commensurable.

There are several ways that this idea might be expressed. A recycling activity must yield a product which has economic value in commerce.
Or, a recycling activity must use fewer resources and produce less waste than not recycling.

Under these definitions of efficiency and appropriate recycling, one might conclude that certain products should not be recycled but converted to energy and/or taken to disposal in the most efficient manner possible.

<u>On recycling goals:</u> <u>Today there is a lot of talk about establishing recycling goals and objectives without an in-depth analysis of the tradeoffs required to achieve those goals</u>.

I argue that goals should not be established in the absence of understanding the costs in other resources to meet those goals.

There is also discussion about defining the meaning of:

- a recycled product
- a recyclable material.

First, anything is recyclable with enough input of other resources; however, it may be inefficient to do so. So, the definition of recyclability must finally relate to efficiency and, I would argue, economic efficiency.

The issues associated with defining recycled content are also extremely complex. It may be that a specific paper product can utilize a small amount of recycled material without dramatically affecting technical performance or cost; yet at a higher level of recycled content, the technical specifications may be dramatically degraded and the costs dramatically increased.

The marketplace sorts out these issues, considering technical requirements, processing parameters, and economics. There is simply no way that an a priori political process can do as well. Optimal use of recycled material will vary with materials and with applications.

Our objective should be to support policies and directions which are efficient. We should be highly skeptical of inefficient approaches.

Lest you think I am against recycling, let me disabuse you of that notion. As producers, we recycle products where it is efficient to do so. We recycle because we try to be efficient of cost factors like materials and energy. That's one of the driving forces in a market economy.

I would like to review with you a classic example of how the marketplace works to achieve efficiency.

All of you recognize the package I am holding as the 12-oz. carbonated beverage can used by the soft drink and brewing industries. When I joined The Coca-Cola Company it was a passage of rites when a young man could squeeze the can and then crimp it with one hand. I think we must have viewed it as a surrogate expression of male virility.

In 1961, the 12-oz. beverage can weighed 164 lbs/1000. The steel industry developed a new technology called double cold reduction, reducing the weight of the steel from 80 - 85 lbs. per base box down to 55 lbs. per base box, thus reducing the amount of steel used in the can. This new rolling technology also produced a smoother metal, reducing the amount of tin required to cover the hills and valleys in the surface. So tin was reduced from 1 lb. per base box to 1/4 lb. per base box.

But there was still that soldered side seam, with the solder being an alloy of tin and lead. As a customer we did not like the soldered side seam, and we told our suppliers so. Continental Can developed a welded side seam, and American Can developed an adhesive bonded side seam ~ and the lead solder was eliminated. Since the can was no longer soldered, there was no need for the tin ~ and the can industry moved away from tinplate to black plate.

The aluminum industry decided to compete for a share of the 12-oz. beverage can market, and they produced a two-piece aluminum can.

And the steel industry responded with the development of a two-piece steel can and both industries, through metallurgical innovations, design changes, and process improvements, continued to innovate and reduce the weight of their respective beverage containers.

Today the 12-oz. aluminum beverage can weighs only 35 lbs/1000, for a reduction in weight of about 80% during my tenure with The Coca-Cola Company ~~ and I can squeeze and crimp it with two fingers ... [TEAR THE CAN] ... and rip the can apart.

Competition between steel and aluminum, between glass and plastic, between cans and bottles, drove the innovation and creativity, and the new capital investments that I have just described.

That innovation occurred in a market economy which forced competitors to add value by being innovative and, in so doing, they produced lighter weight packages with less waste and lower environmental impacts. Government regulation could never have accomplished what the market accomplished. Government regulation would have stalled innovation, fixed technology, and locked in inefficient systems, as centralized planning has done in Eastern Europe.

Now let me switch gears and talk about what I call state-specific packaging. States are doing their own thing. And states are pursuing public policy initiatives that would, in many cases, make packaging state specific.

Laws that make packaging state-specific would create huge problems for the consumer goods industry and its suppliers, its customers, and for consumers.

State-specific laws would ...

1) Increase the number of stockkeeping units which, in turn

2) Would increase inventories which, in turn

3) Would increase the need for warehousing space which would, in turn

4) Increase both working and fixed capital requirements.

5) As stockkeeping units increased, production runs would shorten, down time would increase, forecasting would be substantially complicated, and out-of-code problems for short-life products like yogurt and other dairy products, juices and juice-drinks, would increase and waste would increase.

Suppliers would be required to produce more different types of packages, some with special symbols, some with statements like 5¢ refund, Maine only, some with a minimum recycle content, and on, and on.

And again, for suppliers, stockkeeping units would increase, production runs would decrease in length, down time would increase, and costs would increase. And all levels of distribution would be forced to inventory more SKU's and maintain much larger inventories.

If this country moves to a plethora of local laws and regulations that make packaging state-specific, it will have a dramatic increase on the cost of producing and distributing beverages, food, and consumer products in this country, and it would generate more waste. State-specific packaging either as a result of labeling or material specification, would do great damage to interstate commerce.

I am certain that is bad public policy, yet that is exactly what we are facing.

Now let me switch gears again and talk on another level which I am going to call "Science and Symbols." Determining what is environmentally friendly is not an easy task. It requires a complex systems analysis, from extraction of materials to disposal. It requires studying the domains of air emissions, energy consumption, water effluence, material consumed, etc.

Yet, there appear to be people who can recognize gods and devils without doing this kind of science. At one time in our country's history there was a community in which people could recognize witches. Today we live in a time when people, often without good analyses, attempt to label products as gods or devils.

We live in a world of instant communications. We live in a visual world, and a world where ideas are expressed in sound bites.

The picture of a dead elephant on the Great Serengeti Plain of Kenya with its tusks hacked out of bloody sockets, is a powerful metaphor and symbol of environmental exploitation.   So, in this new world, you must be able to deal with symbols as well as with science.

Unfortunately, industry often finds it difficult to deal with symbols and the powerful metaphor. Our arguments are often complex and require a technical understanding.   We tend to base our arguments on facts, logic, science, technology and economics.   That's a little bit more than a sound bite.

Let me describe an example of a situation with which I am dealing.

I know all of you are familiar with the little square beverage container that we call the drink box.

The package is used primarily by children. It is safe, lightweight, easy to handle, easy to carry, and it protects the product extremely well. Last year the Institute of Food Technologists voted the aseptic package as the most important innovation in food technology over the last fifty years. Yet, the State of Maine recently banned this package because, in the view of the Maine Natural Resource Council and some legislators, it was not recyclable.

In their view, it was not recyclable because it was made from several layers of different materials. That is simply incorrect. In fact, plant scrap from the production of aseptic packages is routinely recycled.

Well, let me give you some bottom-line facts. The package uses less material, less energy, and produces less waste than any alternative single-serving package throughout its production and distribution.

It even does better than all multi-serve alternatives.
It does not require refrigeration like many juice
packages. It truly has an impeccable set
of environmental credentials. The multi-layer structure
of the aseptic package is the reason the package is so
light, and material and energy efficient.

In the case of the aseptic drink box as a first condition,
it uses resources very efficiently, yet that was not
recognized or at least acknowledged by those
advocating regulations in the state of Maine.

My personal belief is that the aseptic package was
selected to be a symbol or social educator for recycling.
We have seen other symbols that have been selected
to drive home a point of view: McDonald's clam shell
packaging is a good example. Some of you may be
producing products that may become symbols in
the decade of the '90's.

Do you know enough about your product, or are you sufficiently prepared when someone describes your product as a devil incarnate?   Well, you need to be. In the case of the drink box, we had been fortunate to commission a study to evaluate the environmental impacts from extraction of raw materials, through every process and transport step, to disposal.   We were at least forearmed when one of our products became a cause.

In the decade of the '90's, I think many of you will be required to participate in public policy discussions regarding the regulation of your products or industries. It will require developing an understanding of intellectual ideas that underlie the environmental debate over environmental regulations.

It is a different activity than producing and selling a product.   Is your company prepared?

Earlier I mentioned mandatory deposit and refund schemes that have been established in some states to aggregate and sort the materials which have been disaggregated and dispersed as part of the consumption process.

Well, those programs will, in fact, accomplish those objectives but at a huge cost. If one does a cash flow analysis on the beverage container mandatory refund scheme like the one in the state of Maine, one will find that the incremental cost for using aluminum for a 12-oz. beverage can is about $1200 per ton (at a 65% return rate), due to the increased handling and transfer activities implicit in the scheme.

If the aseptic package that I just described were placed under a similar scheme, the increased cost for using material in that application would be about $3400 per ton (at a 70% return rate).

Those costs must, in fact, be passed on to the consumer in the form of higher prices. The actual costs per ton vary dramatically based on container weight.

If a state, in addition, wanted to collect for containers not refunded, then there is an additional approximate $1200 per ton associated with what is, in fact, a hidden tax.

Now I would like to switch gears for the last time to talk about public policy, and offer a point of view based on the examples that I've used in this talk.

Good public policy should be based on sound science and economics, not symbols. I am greatly concerned that many of the public policy proposals are symbol related.

Good public policy should be efficient, both economically and environmentally. Many of the proposals being tossed about today would increase environmental impacts and economic costs simultaneously.

Recycling is desirable and virtuous when an economic product is produced, and when the tradeoff with environmental impacts associated with the recycling operation is not negative.

The marketplace works. The marketplace works because competition forces efficiency. Public policy that limits competition or tries to make a priori decisions about what material, or what product, or what package should exist in the marketplace, will, I argue, be very detrimental to our economy and our society.

It's the competition between alternatives that fosters innovation and creativity and improvement, and both economic and environmental efficiency.

I argue that good public policy is broad and should apply to all products which create an externality or social cost. Good public policy is not inefficient as state-specific packaging or mandatory refund schemes are. Good public policy will achieve its professed objectives.

If the issue that is driving us toward package regulation is, in fact. solid waste, then I argue that we ought to attack solid waste in a very direct manner based on the imperatives of each locality or region where waste is produced.

We will never solve the solid waste problem by dealing with one or two products, or one or two materials.

We will not solve the solid waste problem by attempting to move used materials back through a high-cost distribution system.

We will only solve the solid waste problem by recognizing that we will always have to deal with some residual materials, and we must plan to deal with those materials in the most cost and energy efficient manner possible.   That will include municipal recycling of some materials, but taking other materials to incineration and/or disposal with the least amount of additional energy and economic input as possible.

And good public policy would not interfere with interstate commerce.

In closing, I would like to thank Jerry for extending to me this invitation to participate in your conference.
I note that many of the issues I raised will be discussed in more detail by subsequent speakers.

I hope that I have cast my remarks in such a way as to stimulate further discussion.

Thank you very much.

SESSION 3

HOW FOOD COMPANIES USE PLASTICS

Brian L. Popelsky   *Daryl Brewster*

CAMPBELL SOUP AND PLASTICS - WHAT'S NEXT

Brian Popelsky
Senior Manager - Packaging

Daryl Brewster
Business Director
Campbell Soup Company
Campbell Place
Camden, NJ 08103-1799

Biographies

Brian is Senior Manager, Packaging for the Soup Sector encompassing 1.1 billion in sales at Campbell Soup Company. He is responsible for all packaging components for Red and White condensed soups, ready to serve WFT soups and broths, dry and microwaveable soups. He holds a BSME from Rensselaer Polytechnic Institute.

Daryl is presently Business Director for the Soup Sector encompassing 1.1 billion in sales at Campbell Soup Company. Prior to that he was Category Manager for shelf stable foods at General Foods. Previously he was Product Manager for Jell-O Gelatin, Minute Rice and Stove Stop. He holds an MBA from the University of North Carolina at Chapel Hill.

Abstract

"We at Campbell Soup Company, would like to take this opportunity to address the FoodPlas Conference on the State of Plastics as it relates to our business.

An historical perspective covering our company's earlier business and later developments will be presented by Mr. Brewster as an attempt to share how we have fared in plastics food packaging.

We will present national demographics and discuss their relevance and impact upon our business.

History has proven that over 70% of all new products will die in their first year of life. Nearly 90% fail within the first four years and an astonishingly low amount of about 2% become national brands. The cost of failure can be extremely high!! Not only are there the obvious costs in dollars...many millions in fact...but the cost in manpower, marketing momentum and supplier confidence can be nothing short of cataclysmic for our business. We will therefore present various examples of plastics programs at Campbell's and discuss why they were successful and why they failed.

Building on what we have learned in the past, Mr. Popelsky will then present our feelings regarding what is required in a plastic container for the future. Topics such as cost, package performance, decoration needs, product safety, shelf life, and of course environmental impact will be discussed.

160

WHAT'S IT GOING TO BE: FLEXIBLE, RIGID, MICROWAVEABLE, DUAL OVENABLE, OR WHAT?

Bill Lyman
Vice President R & D
Grace Culinary Systems
7911 Braygreen Road
Laurel, MD 20707

Biography

Mr. Lyman attended the Culinary Institute of America and the Dunwoody Institute of baking before serving his apprenticeship with Chef Joseph Chatreau at the Minikahda Country Club. He began his professional career as the Executive Chef at the Wayzata Country Club.

He has served as a research and product development Chef at General Mills Research Center. He also was the Executive Chef with Litton Microwave Oven Company. In 1984 he was a member of the Midwest Chefs Society Culinary Team at the Culinary Olympics in Frankfurt, Germany.

While working at the Cryovac division of W.R. Grace he made important contributions to the Capkold Cook Chill System and the Sous Vide Method for preparing chilled fresh foods.

He is currently Vice President of Research and Development for Grace Culinary Systems. Grace Culinary Systems produces and distributes fresh soup, sauces, entrees, composed salads and bakery products from their facility in Laurel, Maryland to 17 American Cafe Restaurants located in the Washington-Baltimore area and other Foodservice accounts on the east coast.

Abstract

A look at how a new company is custom preparing fresh chilled, sous vide and frozen foods for supermarkets, food service accounts and the selection process for the type of package requested by the customer.

WHAT'S IT GOING TO BE:
FLEXIBLE, RIGID, MICROWAVEABLE, DUAL OVENABLE OR WHAT?

I   INTRODUCTION

     o   Description of Grace Culinary Systems

II   SIX CATEGORIES OF PLASTIC PACKAGING USED AT GRACE CULINARY SYSTEMS

     o   Kettle Produced
     o   Horizontal Rollstock
     o   Bulk Salads
     o   Retail Packages
     o   Shrink Wrapped Product
     o   Miscellaneous

III   SUMMARY

     o   Foodservice
     o   Retail

I. INTRODUCTION

HOW FOOD COMPANIES USE PLASTICS?

What's it going to be: Flexible, Rigid, Microwaveable, Dual Ovenable or What?

The title to this presentation comes from the fact that our food company uses all of these plastics in our operation. Yes, we use flexible packaging for bulk and packed products. We use rigid plastic containers for prepared products and microwaveable plastics for retail packed entrees. We use dual ovenable plastics for another retail food product and we also use a bright shiny product called Aluminum, or AL-U-MIN-E-UM if you will, for another product line for the retail market.

But first of what you need to know is that Grace Culinary Systems is a new food company. We are located in Laurel, Maryland in a 65,000 square foot facility that we moved into in May of 1989. We are owned by the W. R. Grace & Company whose company headquarters are in new York and will be until this summer when it will move to Florida. Many of you probably already know this but I have found that our logo design and name are not associated with W. R. Grace by many people that I have talked with for the first time.

Grace Culinary Systems food production experience is actually seven years old and is the central food production facility for the American Cafe restaurants. There are 17 American Cafe restaurants located in Washington, D.C., Virginia and Baltimore and our first franchise restaurant opened in Richmond, Virginia in November of 1990. The first American Cafe opened in Georgetown and from the beginning it was the plan to prepare foods at this facility for this location and to open other restaurants and supply them with the same top quality foods as served in Georgetown. As restaurants were added, it was necessary to move the central production to a separate facility in Brentwood, Maryland where it operated for four years until moving into our existing facility in Laurel, Maryland. The new facility was designed to be a USDA inspected central production facility and bakery for the American Cafe restaurants and also to produce top quality food products for other foodservice accounts such as hotels, restaurants, health care, business and industry, supermarket deli's and convenience stores. The original plan was to make available similar products as sold at American Cafe restaurants with the same emphasis on quality restaurant food.

Grace Culinary Systems employs approximately 100 people. We have 40 food production employees, 22 bakery employees, five quality assurance employees, and four in research and new product development. The production plant currently operates five days a week, one shift on food production and the bakery currently operates three shifts a day, six days a week. The bakery schedule will change when the two additional ovens that are being installed are operating.

165

Now, to address the title of this segment of how our company uses plastics. But before we get into that, you must also know that Grace Culinary Systems owned by W. R. Grace also owns the Cryovac Company and I spent 12 years with the Cryovac Division of W. R. Grace before transferring to Grace Culinary Systems and there may be some bias towards flexible plastic packaging because of that. You can be the judges.

II SIX CATEGORIES OF PLASTIC PACKAGING

Grace Culinary Systems has many plastic packaging systems that I have divided into six categories:

1. The first category is a kettle system for soups, sauces and hot fill pumpable main dish meals and cold mixed dressings and sauces. This system was developed by Cryovac and is sometimes referred to as capkold. Capkold is a name used by one equipment company and refers to "Controlled Atmosphere Packaging Cold." From specially designed kettles, we cook and mix items and when properly cooked and the Quality Assurance Department has verified the viscosity, color and flavor, the hot product is pumped at 190°F into a flexible plastic casing. The casing is clipped closed with a label affixed to it with the product identity, ingredient statement, production date and use by

date. The casing is then placed into a tumble chiller and immersed in chilled (approx. 35°F) water and tumbled for 30-40 minutes. The product is cooled from the 190°F to less than 40°F. The chilled product is stored at 38°F and will last for 3-4 weeks. Cold dressings and sauces are pumped into the plastic casings, labelled, closed, and stored at 38°F also. We produce over 25 different soups, 15 different sauces and several entrees such as chicken pot pie, beef pot pie and many concentrated sauces and dressings. These products are for the American Cafe restaurants and other foodservice accounts. We pack these products in various sizes, from 1 quart to 2 gallons. Normally, we market them in 2 - 1 gallon units per case or 4 - 1/2 gallon units per case. Cryovac has since added a more automated vertical form fill and seal system with a continuous chiller. This new system has been gaining good acceptance.

2. Our second method of plastic packaging is a horizontal roll stock machine from Koch Multivac. On this machine, we pack some raw products like chicken taco filling, beef stroganoff, Italian meatballs, sausage and peppers, chicken couscous, meatloaf and others into three to 10 pound units for cooking. We also pack individual portions of raw chicken, seafood, beef, pork and lamb for cooking. All of these products are cooked and cooled in our steriflow cooker. A specially designed waterproof label is placed on each package with

a product name, ingredient listing, production date, and a use by date. These products have chilled shelf lives from five days for some up to four weeks for others. We also freeze some of these products. This is "Sous Vide".

On this same Multivac, we also package five pound units of composed salads like chicken salad, chicken tarragon, tuna salad and others. We have four sets of dies for the Multivac, so we can make most any practical size we may need for our customers. Some of the other products we package in five lb. units are fresh mashed potatoes, artichoke dip, hot and spicy buffalo wings, barbequed ribs, herb rice, and others.

Using the same Multivac for another plastic package, we pre-portion some chilled sauces such as hollandaise, bearnaise, Alfredo, Russian mustard, lemon leek, red clam, and white clam sauce in two ounce to eight ounce portions for freezing. The film we use on this machine is obviously a laminate that is freezable and boilable and I don't have to tell you who the supplier is. This pack size is very convenient for hotels and restaurants and allows them to have a large variety of sauces on hand to accompany pasta or meats that they can heat in less than 10 minutes from the freezer without any waste.

3. The third category of flexible plastic packaging we have is for large volume, less expensive, bulk salads such as cole slaw and macaroni salad. It is also used for cooked pasta, such as linguini and egg noodles. The salads are mixed in our chilled salad area in large batches then delivered to the packaging room. The pasta is cooked and chilled rapidly and also delivered to the packaging room. These items are hand filled into a barrier type bag, clipped and labelled. The 7-1/2 pound bags are then either cased or distributed on our rack system to the American Cafe restaurants.

4. The fourth category of plastic packaging is a rigid container by our definition. We use several rigid containers in our facility and in various shapes. Some are dual ovenable and some microwaveable only.

One of our retail accounts has us pre-portion several salads for them in pint containers and 1/2 pint containers. We pre-package chicken salad, cole slaw, macaroni salad, chicken tarragon, tuna salad, and turkey gravy. These containers are labelled and pre-priced by us with the appropriate UPC labeling. We mix the products, package them in bulk or portions and ship all on the same day. The products are delivered to a central distribution center and delivered to the retail stores the next morning. The stores only have to open the cartons and place the product in the chilled case for sale.

Another rigid plastic container we use is dual ovenable and used for products like stuffed potatoes, chili con carne, tuna bake, beef burgundy and items like these. A snap-on lid is put on the container and then the container is covered with a colored sleeve that has the label information, heating instructions and UPC code. It makes a very merchandiseable package and is distributed the same way as the portioned salads. The instructions on some of these products to the consumer tell them to "use within one day or freeze".

One of the most attractive rigid plastic containers we use is for quiche. We produce three types of quiche; broccoli and cheese, spinach and cheese, and bacon and cheese. These are all fully baked from raw in an aluminum pan. They are chilled thoroughly to below forty degrees, then placed in a custom made rigid plastic container and covered with a clear plastic see-thru dome. We shrink wrap this to make it tamper evident and apply the appropriate labels.

During the holidays, we pack two pounds of buffalo wings in a dual ovenable container. We place a snap-on lid on it, shrink wrap and label it for distribution. It makes a good looking package.

Beef pot pie, chicken pot pie and turkey pot pie also take advantage of rigid plastic packaging. These are fully baked in foil pans with a puff pastry crust, rapidly chilled and placed in a clear large deep plastic container with a snap on lid. We put a sleeve around it with the proper label. Heating instructions are for a regular oven only. Another unique item like this is a vegetable and potato bake. The aluminum container is used for the product because the reheat in a regular oven is necessary to achieve proper crust quality and some browning. We have two more products like this in development that are meat based but use mashed potatoes on top with shredded cheese instead of puff pastry.

5. The fifth category we utilize for plastic packaging is shrink film packaging. We use shrink film in many ways. We produce pre-cut New York style cheesecake for foodservice sales and also for large warehouse style supermarkets. After the frozen cakes are cut and the slices separated with parchment paper, we shrink wrap the whole cake to a square board. This holds the cake together and keeps it in the center of the square board. It is then put into a box and sealed with tamper evident tape.

We also produce individual quiche; turnovers filled with ground meat, barbecue chicken, spicy chicken or vegetables and cheese; and croissants filled with ham and cheese or spinach and cheese. We place these on a board and shrink wrap them to protect them in transit.

Blueberry muffins, raspberry muffins, banana muffins, and oat bran
muffins are also shrink wrapped before freezing.

6. The sixth category is somewhat of a catch-all for the miscellaneous
   areas where less sophisticated plastic material is used. For example,
   we supply our American Cafe restaurants with whole wheat, sourdough,
   rye and caraway cheese breads sliced and packaged in poly bags.
   All-butter croissants, cherry, apple, raspberry and cheese danish are
   overwrapped with plastic bags for subsequent freezing and shipping to
   the American Cafes. Chocolate chip cookies, peanut butter cookies and
   coco loco cookies are packed in polyethylene bags and boxed for
   shipment. Sheet pan racks with product in pans are covered with large
   plastic bags for storage and transportation in our plant. All of our
   trash containers are lined with plastic and we stretch wrap our
   pallets of boxed product to prevent shifting while being transported.

III SUMMARY

By now perhaps you have come to realized as I did the title, "What's It
Going To Be?": Flexible, Rigid, Microwaveable, Dual Ovenable or What? fits
for Grace Culinary Systems.

The kettle system and the multivac systems both use flexible plastic and are very essential to our business of supplying the foodservice market and bulk foods for retail. They are, however, not attractive retail packages. The food quality inside these packages in our opinion is excellent, and our food service customers who use our products are very pleased also. The rigid plastics we use for the salads, quiches, pot pies and potato skins really do a great job of merchandising the foods and our customers who sell these products are very pleased with our consistent quality. As long as we continue to deliver quality products to our customers in the appropriate plastic package, we feel we will succeed and grow.

FLEXIBLE PACKAGING AND THE ENVIRONMENT

Curtis W. Babb
Manager
Packaging Research
Hershey Foods Corporation
Hershey Foods Technical Center
1025 Reese Ave.
Hershey, PA 17033

Biography

Fifteen years experience in food and medical device packaging at Hunt Wesson Foods, Baxter Travenol Laboratories, Hershey Foods Corporation. B.S. in Packaging from Michigan State University 1975, MBA Drexel University, Philadelphia 1990, Member of ASTM, NFPA, Institute of Packaging Professionals.

Abstract

Consumers and customers are demanding environmentally friendly packaging. Recycling of films needs to begin. Monolayered films need to be developed to replace laminates. Post consumer films need to be used in secondary and primary packages.

FLEXIBLE PACKAGING AND THE ENVIRONMENT

Presented by

Curtis W. Babb
Manager, Packaging Research
Hershey Foods Corporation

March 6, 1991

Introduction:

I accepted this invitation to speak to you today because I have a vision of how the future could be. This vision is your opportunity to survive the present environmental challenge and even thrive in the future. This is my vision.

I see the food and packaging industries making a positive impact on the solid waste crisis. We will be recycling flexible films. We will recycle or replace high barrier laminations. We will develop high barrier monolayer films. We will use post consumer packaging material in secondary packages. We will use post consumer packaging material in primary packages. This vision can become a reality with your help. You should have a clearer understanding of each of these opportunities by the end of this presentation.

Plastics applications at Hershey Foods Corporation will first be presented. Their functions, benefits, and problems will be summarized. Next, examples of feedback from our customers and final consumers will be highlighted. Letters, telephone calls, and focus groups provide a steady flow of information. Finally, our future needs will be projected. History will judge you and I by how we respond to these needs. My vision can become our vision.

Plastics at Hershey:

Plastics are used to package a number of our products. Hershey's Chocolate Syrup in a blowmolded squeeze bottle and Hershey's Cocoa in an injection molded tub are familiar to most of you. Many of our individual candy bars are wrapped in laminations of plastic, paper, or metal. This presentation will discuss three product applications in some detail.

Hershey's Bar None is a sandwich of wafers and chocolate cream topped with nuts and enrobbed in chocolate. The product water activity is much lower than normal room relative humidity. This creates a tendency for the product to pick up moisture from the environment. When this happens the product loses its desired texture and becomes soggy and tough. The metalized polyester lamination provides the needed moisture barrier for a reasonable product shelf life. The structure also complies with FDA regulations for food contact materials, provides insect infestation resistance, a degree of tamper evidence, good machinability on our production lines, all at a reasonable material cost.

Hershey Pasta Group is a major supplier of branded pasta products. Our Light 'n Fluffy, American Beauty, Skinner, and Ronzoni brands use laminations or coextrusions of polypropylene and polyethylene to bag many of their products. Material clarity, coefficient of friction, puncture resistance, stiffness, hot tack, and printing quality are all important performance characteristics. Bond strength between the plies need to be increased and costs improved. These dry pasta applications are not as likely to extract substances from polymers and therefore have less severe requirements from the FDA. This characteristic offers some opportunities to be detailed later in this presentation.

Hershey's Chocolate Bar Flavor Puddings were introduced into many areas of the country in 1990. This product is filled into an injection molded polypropylene copolymer cup. The lidding is a lamination containing foil and plastic. This package withstands hydrogen peroxide and heat sterilization, refrigerated distribution, provides product visibility, tamper evidence, and runs well on high speed equipment.

The Society of Plastics Industry recommended code has been added to most of Hershey's rigid plastic containers. The need exists for an identification code to be used on flexible plastic packages. This is clear from many comments we have received from our consumers which will be presented next.

Customer and Consumer Feedback:

Hershey has received letters from its customers asking which of our products are environmentally friendly. Some of the larger customers have created special sections in their stores to feature packaged products that are environmentally friendly. They also ask what are we doing as a company to deal with the solid waste crisis, what programs do we have in place, what changes are we making.

Our consumers write to us or use our 800 number to ask us some of these same questions. Is our packaging biodegradable? Is our packaging able to be recycled? Is our packaging made from recycled material? How do you suggest we recycle this package? They are also quite free with their comments.

"Shame on you for using a plastic can."
"Where do you stand on Oregon recycling ballet 6."
"Use recycled paper."
"Won't buy plastic."
"Trouble recycling container."
"Use glass instead of plastic."
"Use paper cans."
"Use aluminum cans."

These people are not telling me to use plastic. We do get a few positive comments about the <u>functionality</u> of our plastic containers but no acknowledgement that it is made from plastic. We do our best to answer their questions.

We tell our customers and consumers about our efforts to recycle waste generated in our offices and our plants. We tell them about the large amounts of corrugated fibre board we use that contains 25% recycled fibre. We tell them about our folding cartons that are recyclable and that many of them are made from recycled newspaper. We also point out that our rigid plastic containers have the SPI recycling codes on the bottom of each container.

177

But our answers and actions simply are not enough to satisfy them. As a result, we are ready to begin some programs to change some of our packaging to make them more compatible with the environment. A number of needs have been identified which compose the vision shared with you today. These needs will be presented next.

Future Needs:

We need to recycle flexible films. The plastics industry has done an outstanding job of making the recycling of rigid plastic containers a reality. You received help from the bottle deposit states which provided a source for recycled polyester. States with mandatory curb side sorting trash laws helped provide a source for recycled high density polyethylene. Companies like Proctor and Gamble, Rubbermaid, Unilever, and the bottled motor oil companies are using post consumer plastics in their packaging. The Environmental Protection Agency recommended source reduction, recycling, incineration, and landfilling as the strategies to be followed by industry in reducing solid waste. As a result, flexible packages will be used as a source reduction. This will increase the amount of flexible materials to be disposed of. Incineration adds more carbon dioxide to the atmosphere and the green house affect. Landfilling is not the answer anymore. Recycling flexible films will be targeted next by the environmentalists. Litter on land and sea has driven the demand for biodegradable materials. Recycling flexible films will adequately dispose of this problem.

We need to develop and use monolayer films that are recyclable. Single material films will be more recyclable. They need to be identifiable. The Flexible Packaging Institute needs to come up with an identification system or their customers will. Wrapping equipment needs to be designed to handle these single material films at the same production speeds used for multiple material laminations and coextrusions. The collection infrastructure needs to be developed to handle films. If possible it should be compatible with the rigid plastic system or even become an integral part of the same system.

A high barrier single material film is needed to replace metalized polyester and foil laminations. These present laminations could be used in commingled products but even these products eventually end up in the waste stream. If metalized laminations are banned then companies with sensitive products will be forced to use more layers of single materials in secondary packaging. The end result will be more solid waste, not less.

Post consumer flexible materials need to be used in secondary packaging. Perhaps another market for recycled rigid plastics is films for non-food contact applications. Hershey uses films as overwrap on six pack trays and bags for smaller products. This would be an excellent application for recycled films.

Post consumer materials need to be used in primary packaging. The recycled material could be cleaned and then buried between virgin layers of the same material. Some dried products may be compatible with direct food contact of recycled materials. Pasta is packaged in clay coated newsback. This material contains recycled newspapers and complies with the Code of Federal Regulations for use with this food product.

Even if all these needs are met, what will be next? Will it be possible to establish closed loop recycling for all materials? Paper experiences a 30% reduction in strength each time it is recycled. The thermal history of polymers affects its next use. The use of secondary uses for recycled materials will only delay the ultimate trip to the landfill. Closed loop recycling of materials somewhere in their life cycle needs to be established or the solid waste problem will just be postponed for your children to deal with.

Conclusions:

We are all part of the solid waste problem. We all consume packaged products. We all dispose of the package and product when we are finished with it. We all can be part of the solution to the solid waste problem. As consumers we can recycle our discarded packages and products. As business people we can recycle our office and manufacturing waste. As packaging professionals we can do even more.

Flexible films need to be recycled.

The future needs or vision I and others have shared with you today may become a road map to dealing with this environmental issue. The choice is ours.

SOLID WASTE SOLUTIONS FOR FOOD AND BEVERAGE PACKAGES

Dr. John Rotruck
Director of Research and Development
Food and Beverage Business
Procter & Gamble
Winton Hill Technical Center
6250 Center Hill Road
Cincinnati, OH 45224

Biography

Dr. John Rotruck received his PhD and MS in Biochemistry from the University of Wisconsin in 1971. Since then he has been with Procter & Gamble and is currently Director of Research and Development for the food and beverage business for which he has responsibility for managing safety, regulatory and environmental efforts. He is a member of The American Institute of Nutrition and is active in a number of professional and trade associations including ILSI.

Abstract

Procter and Gamble's approach to solid waste and its application to food and beverage products will be discussed. The integrated waste management approach has been used to measure solid waste environmental attributes and to develop strategies for future packaging changes. Key issues facing food and beverage packages will also be discussed.

Thank you very much and good morning. I am pleased to be with you and to have the privilege of sharing a few thoughts with you. I, of course, want to discuss certain aspects of the solid waste issue. In particular, as my title suggests, I want to talk about P&G's approach to managing solid waste, in particular as relates to plastic and the solid waste crisis. While it certainly seems as if we are dealing with a major problem and perhaps even a crisis, I have come to view this problem as an opportunity and thus, my focus on solutions. While I realize the audience is very interested in plastics, I do not want to just consider plastics simply because I do not think it appropriate to consider only one material or one type of packaging. Unfortunately, all too frequently a product or package is identified as the problem and becomes the focus of legislation or package bans.

My approach will be to briefly give some overall perspective on garbage, including a brief but complete history. Then I will discuss Procter & Gamble's overall policy and program on solid waste issues and finally discuss how we are using these principles in the food & beverage business.

Now, a few words about the discovery of garbage and its history. It all started in 400 BC when the Greeks developed the first town garbage dump. I would also indicate that refuse was a problem for prehistoric tribes and the native tribes of North America who actually moved away from this garbage.

In 200 AD, the Romans thought up the brilliant idea of having sanitation workers. Two men would follow a cart and throw the garbage they found into it. Obviously, there's been little improvement on this idea in 1800 years. One of the few times garbage figured into medieval history was during a 1415 Portuguese attack on the Moroccan city of Ceuta. Two princes charged up what appeared to be a strategic point only to discover they had heroically captured the garbage dump.

In 1551, the first recorded example of packaging appeared. Andreas Bernhart, a German paper maker, began placing his paper in a wrapper bearing a design and his name and address. The first municipal refuse-collection system was started by Benjamin Franklin in 1757. This was a very innovative system which avoided the need for landfill by having the slaves who collected the garbage wade into the Delaware River and toss it into the current. 1776 was a landmark year in the history of garbage. In that year, we saw the first act of recycling in America. Patriots in New York knocked down a statue of King George III and remade it into 42,088 bullets. In 1834, the first solid waste legislation was passed. Charleston, West Va. enacted a law that protected vultures from hunters because they were critical in keeping the city's streets and yards clean from garbage.

The plastics industry was born in 1869 when American John Hyatt made "celluloid."

The first instance of "NIMBY" (Not In My Back-Yard) appeared to be in 1894 when, to protest their up-river Washington's practice of using barges to float their garbage down the Potomac, the good citizen's of Alexandria, Va. began sinking the barges before they reached Alexandria.

181

In 1935, we see the birth of the <u>can</u> of beer. In 1938, the Xerox process was invented by Chester Carlson, a 32-yr old Physicist. Styrofoam is invented by Dow in 1944. A single-day garbage record was recorded for New York in 1945 when 6,334 tons of trash was dumped in the streets for V-J Day, end of WW II.

Moving along to 1962, Rachel Carson published "<u>Silent Spring</u>." This warned us of pesticide use and, more importantly, reminded us that when you throw something away, it doesn't really go away.

In 1968, the aluminum industry began aluminum recycling. In the 20 years since that time, 10.5 billion pounds have been recycled saving 60 million kilowatt hours and enriching collectors $3.75 billion. In 1973, a Harris Poll was recorded indicating who the American people really trust. 52% felt they could trust their garbage collector; only 18% felt they could trust their President. And we all remember the wonderful 4-day historic tribute we paid to the Statue of Liberty in 1986. We left behind 2,079 tons of garbage. With the 1987 sailing of the garbage barge "Mobro" with its 3,100 tons of Islip garbage, we finally realized we had a problem and this brings up up to present.

As you have probably already heard, we're seeing more legislation than we can keep track of. Last year, we had more than 1,000 state and local bills on solid waste. They year it will be more. Also, as you have probably heard, our loss of landfills is one of the primary reasons for the solid waste crisis. In 1985, we had more than 5,500 landfills in this country and by the year 2000, we're going to have less than 1,500. Landfills are filling up and they're closing them at a rapid rate and new ones can't be easily sited.

Clearly awareness of this issue is growing at a huge rate. In 1980, only about 10% of the German people cared about the environment. By 1988 environmental protection was just as high of a concern to the German people as unemployment, the highest concern in the country. We're seeing the same growth of concern in this country. We're just a few years behind Europe, that's all. It's growing very, very fast, and I think we can expect the same kind of growth of concern in this country.

So, let me tell you just a couple of important facts. Each American generates about 3.6 pounds of municipal solid waste a day. And the way this is fractionated in the waste stream, about 36% of it is paper and paperboard. It is important to note that about 20% is yard waste -- your grass clippings, and your yard waste. And about 9% is food waste. Those things don't even have to be in our garbage cans -- they can be composted. Another thing to note is that plastics are only 7% of the waste stream. This is another piece that the American consumer has a little bit out of perspective and needs some education on. That doesn't mean we shouldn't do something about plastic, but it's not the problem that people make it out to be.

If you look at how this fractionates in terms of broad categories, you see that containers and packaging are about 33% of the municipal solid waste stream. And of that 33%, most of it is paper and paperboard, and only 4% is plastic.

I think the bottom line here is that nobody is going to escape. This isn't a problem that can be handled by just one sector of our society. It belongs to all of us. And I think all of us are going to have a part.

In fact, we as a large consumer products manufacturer, feel that environmental quality is a new consumer need. In other words, it is as much a consumer need in products now, as building in better cleaning for detergent powders; better absorbability for diapers. And consumers frequently remind us that the consumer is really king and we had better learn how to meet their needs in this important area of business.

Through our early experience in Germany and Europe, we came to view environmental quality as a new consumer need. Similarly for the food and beverage business, I submit that environmental concerns will become as important as other attributes like taste, nutrition and convenience. Certainly, product safety is critical and packaging must perform its critical role in preserving food safety.

Well, what's the answer? As I was telling you before, the solid waste issue has been addressed by the scientific community, and the experts are beginning to agree on what the solutions are. So, because we have at least the scientific community agreeing, this is an issue on which we can began to act with some kind of confidence that we are doing the right thing. And what the solution is, is something called Integrated Waste Management. Integrated Waste Management is a set of four solutions that we use in a particular order, because the ones at the top are better to use than the ones at the bottom, in terms of the health of our environment. So at the top, we have source reduction. And source reduction of course means, make less of it, use less to make it. Don't make it in the first place. And this is a story we need to do a better job of telling especially relative to plastics. The second one is recycling, reusing, and composting. When we talk about composting, we mean a very specialized system that can convert up to two-thirds of the waste stream to soil-like material. These are all methods by which you can take the waste and return it back to some useful life. The third one is waste-to-energy incineration. I think all the experts agree that we can't take care of the whole waste stream by the first two methods as much as we would like to, and that there will be some need for waste-to-energy incineration in this country. And at the very bottom, there is landfill. In other words, try and do something else with it before we put it in a landfill.

There is a philosophy behind solid waste management, since this issue is really being worked at the local level. The better any city does the first two things on the list, the less they will have to use the bottom two things.

Well, how are we doing now? Currently we landfill more than 80% of our solid waste. And in light of the data that I mentioned on landfill closings, we obviously can't continue to do that. The EPA has set goals for 1992 of reducing the solid waste stream by 25% through source reduction and recycling; increasing incineration to 25% and reducing landfill to 50%. These are interim goals. We obviously would like to get down to where we landfill only about 10%.

Now, I'd like to talk for a few minutes about what kinds of things manufacturers can do to make their products more environmentally acceptable. I'm going to show some examples of products. First, Concentration — concentrating products is a very good way to save on packaging. Downy triple concentrate is a way of saving two-thirds on packaging. Concentrated cleaning powders are coming along. These, as well, save on packaging and save on space. Consumers like this because they don't have to carry home as big a box.

<u>Combination Products</u> -- Being able to put two or three products in the same box is another way of significantly saving on packaging. In other words, Tide with bleach saves the need of a bottle of bleach and a box of detergent. Bold saves a need for a bottle for softener and a box of detergent. Pert Plus is an example of a product in which there are two products in one; you don't need two bottles in your shower, you just need the one.

The refill concept has successfully built our business in Europe. As I said, the European consumer is several years ahead of the American consumer in being concerned about this issue. And the German consumer has really realized what this product offers in terms of saving on solid waste. Lenor is the equivalent of Downy fabric softener in Europe. What this concept involves is that the consumer buys the 4-liter bottle of Downy or Lenor. They take it home and when it's gone they don't throw the bottle away. Instead, they save it and buy this small plastic packet of concentrate. They put it in the used Lenor bottle, fill it with water and have a reconstituted bottle of softener. This saves 85% on packaging.

Obviously, we are very anxious to do this same kind of thing with the American consumer. We have introduced this concept in Canada, in four brands. In the U.S. we have just introduced into test market in the Washington and Baltimore area, this version of the refill concept. In the U.S., we have had to use this carton instead of the pillow pack because the American consumer is not familiar with the pillow packs. In Canada and Europe, milk and other products are sold in pillow packs, so the consumers are used to them. In the U.S. this is an unfamiliar kind of package. So, we had to use something a little more familiar.

The initial results are very encouraging. We've got our fingers crossed and we're hoping that the American consumer will start to want to use things like this very soon.

Another thing that we are doing is getting heavy metals out of all of our inks, dyes, and pigments. This, of course, ensures safety in incineration as it makes the ash non-toxic.

More than 70% of our cartonboard is recycled. We're trying to use recycled materials wherever we can. Spic and Span is now being marketed in a 100% post-consumer recycled PET bottle. This is the kind of plastic that 2-liter pop bottles are made from. And just last spring, we announced that we also now have a technology to be able to use post-consumer recycled plastic in our detergent bottles, which are high density polyethylene (HDPE). The technology here is much more complicated. So right now we can only use about 20-30% post-consumer in these bottles. But we are expecting to be able to improve that technology very rapidly and grow to a higher percentage of use.

We are coding all of our bottles with the SPI code, which will tell recyclers and consumers what kind of resins the bottle is made from. You'll start to see these little embossed symbols on the bottom of plastic bottles and this will be very helpful in getting our recycling systems going.

Not surprisingly this same approach fits food and beverage products. Let me tell you how we approached this issue. Basically, I used a variation of integrated waste management to benchmark our packages because in our business the primary solid waste issue is packaging. I surveyed all of our packaging for 3 attributes: source reduction, recyclability, and recycle content. What we found was that we needed to use a variety of approaches because we use a variety of packaging; including single resin plastics, like PET and HDPE, plastic laminates, metallic laminates, paperboard, corrugated, composite cans, aluminum, tin plated, steel, glass, aseptic boxes.

So what have we done so far? What I found surprised me in that close to 80% of our packages already possess a significant solid waste attribute. Source reduction - we, like many manufacturers have light weighted many of our packages. Frequently using aesthetic pleasing designs like this Tiffany bottle we use for Puritan and Crisco. In addition, we have found that by using different packaging we can significantly reduce the amount of packaging like we did for our Folgers brick pack.

We also found that a number of our packages are already recyclable like these aluminum and tin plated steel cans. In addition, our glass bottles are recyclable and all indications are that our plastic bottles like this Crisco PET and Citrus Hill HDPE jug will be recyclable. In fact, in cooperation with our packaged soap efforts we are planning to put our PET and HDPE food and beverage containers into Spic & Span, Tide, Cheer, and Era. While some people believe there is a glut of potential recycled material, I know in some cases we and others are actually having difficulty getting enough recycled material.

Finally, recycle content. As you know, aluminum, steel and many paper products already contain recycled material. And, where possible, we are making sure our various packages contain recycled material. For recycled paper, it's important to note that this is not in direct contact with the product. However, because aluminum, steel and glass are actually melted, any potential contaminants are removed and recycled material from these sources is used in direct food and beverage contact.

While this is not a complete list of our products, I think you have an idea of how we are approaching our products. Well, you might ask, where do you go from here? More of the same, i.e., source reduction, recyclability, recycle content and ultimately packages will be compostable. But I'm certainly not saying every package needs to have every attribute, but I do believe that most of our packages will have to possess at least one of these attributes.

I would like to conclude by discussing a few of the key issues facing your industry. Certainly plastics recycling and reuse of post-consumer plastics is a huge one. We need to do more to support plastics recycling and reuse of plastics in a safe way. There are already major efforts that some of you may be involved with. Coca-Cola and Pepsi have taken a major step and propose to use a very specialized post-consumer material that has been converted into material that is indistinguishable from virgin plastic. This material has been approved by the FDA.

Laminates - Clearly, we love them in the food and beverage industry. They provide critical barriers for our products. Although they are truly a wonderful way to achieve source reduction, their lack of recyclability will be an issue. I would certainly like to tell you that recyclability of laminates is not a big deal, but a fortune awaits the manufacturer who develops materials that do the job of laminates that are easily recyclable.

In summary, solid waste solutions for our business are really not much different than other business. Indeed, solid waste solutions need to become part of our business. We cannot simply say food and beverage products are different. Packaging preserves our products and make them safe. Sure it does, but we need to make environmental attributes an important part of our business strategies and set about solving these problems.

SESSION 4

REGULATIONS FOR PLASTICS PACKAGING:
WHAT'S COMING?

SOLID WASTE PACKAGING ISSUES: GMA'S PERSPECTIVE

Elizabeth Toni Guarino
Senior Counsel
Regulatory Affairs
Grocery Manufacturers Association
110 Wisconsin Ave., N.W.
Washington, DC 20007

Biography

Elizabeth Toni Guarino is Senior Counsel, Regulatory Affairs for the Grocery Manufacturers of America, Inc. She provides legal advice to the association on a variety of regulatory, legislative, and policy matters relating to FDA, USDA, EPA, FTS and other government agencies. She came to GMA in July 1990 from Kraft General Foods, Inc. where she was Senior Food and Drug Counsel. Prior to her four years at KGF Ms. Guarino served six years in the Advertising Practices Division of the Federal Trade Commission. Her last position there was Food and Drug Program Advisor. Ms. Guarino received her B.A. from the State University of New York and her J.D. from Washington & Lee University School of Law. She is a member of the Virginia, District of Columbia and American Bar Associations.

Abstract

As federal, state and international regulations are enacted to address the solid waste "crisis", the special role packaging plays in preserving the integrity and quality of food (as compared to other consumer products) must be taken into consideration. Neither plastics nor any other individual type of packaging material is, per se, "bad" for the environment. Many factors must be considered by regulators, consumers, and industry in devising regulatory approaches to solid waste management.

<u>THE REGULATION OF PLASTICS PACKAGING: WHAT'S COMING</u>

GMA'S PERSPECTIVE ON THE REGULATION OF FOOD PACKAGING AS SOLID WASTE --

PACKAGING BANS AND RESTRICTIONS ON THE USE OF

ENVIRONMENTAL MARKETING CLAIMS

BY

ELIZABETH TONI GUARINO

SENIOR COUNSEL, REGULATORY AFFAIRS

Prepared for presentation at The Plastics Institute of America's 8th
Annual Conference, FOODPLAS '91, Orlando, Florida, March 5-7, 1991

I am going to give you GMA's perspective on one narrow but highly visible aspect of your general topic, which deals with issues and opportunities for the use of plastics in food packaging. I am further narrowing the scope of my remarks to one aspect of this panel's topic -- the regulatory outlook for plastics packaging. I will focus on the recent regulatory activities involving food packaging as solid waste, and specifically on the issues of proposed bans on the use of certain materials, and controls on the use of environmental marketing claims. Most of these regulatory initiatives are not specific to plastics, and in many cases are not specific to food, as opposed to other types of consumer goods packaging. All of them, however, encompass the use of plastics for packaging food products.

First, let me give you a brief description of GMA, so you will better understand our interest in this regulatory arena. GMA is an 80 year old national trade association of approximately 140 companies that manufacture food and non-food products sold in retail outlets throughout the United States and internationally. GMA member companies produce more than 85% of the packaged food and non-food products sold nationwide, exceeding $280 billion in sales last year. Clearly, our members have a very keen interest in the regulation of packaging for food.

I have been asked to forecast the regulatory future, but I would like to begin with some background information to put the issues in context.

As we all know, within the past year or so, there has been a surge of publicity concerning what has been described as a "solid waste crisis". Now let me read quotations from commentators describing the solid waste problem.

Today, Americans - with only 7 percent of the world's population - consume a third of the world's energy and nearly half of the earth's raw materials. Most of the raw materials end up on the nation's trash heaps. In the past, it seemed to matter little what we threw away or where we threw it. A big country with a small population, we could afford to recklessly squander our resources. Over the years we were intent on converting the wealth of America into an abundance of consumer goods. We applied the best technology and the finest management skills to every step in the production, marketing, and distribution of consumer products. We made these systems the most efficient and economical in the world so that we could choose from an amazing array of goods - then use, discard, and replace them at will. But we forgot to take into account the final step in the process. We failed to apply either modern technology or modern management to

the ultimate disposition of this abundance. The
result of this failure is a solid waste crisis.

General prosperity and affluence, coupled with an
appetite for consumption and a desire for convenience
have led to the creation of a national problem of
immense proportions out of the disposal of cans,
bottles, boxes, papers, garbage and countless other
throw-away items. The problem is complicated by ...
mountains of commercial, office, agricultural, and
industrial wastes. Every year millions of tons of
... material resources are dumped, burned, or buried,
forever altering thousands of acres of land, oft-times
fouling air and polluting water, ... all at enormous
public expense.

You may be interested to learn that these quotes are from reports on
solid waste management prepared for the League of Women Voters and the
St. Lawrence County (New York) Planning Board -- both in 1972.

This point is not meant to diminish the magnitude of the current
solid waste problem in the U.S. It is meant to show that the nation's
present solid waste volume does not constitute a crisis -- which, by
definition, is an event or situation involving "sudden or abrupt change"

-- because it did not occur suddenly or abruptly. It is, rather, a genuine problem that has developed gradually and entirely predictably over time.

In fact, since the early 1970's when the country was in its first phase of "environmental awareness", much has changed. Legislation has been enacted to improve air and water quality and solid waste management techniques have improved significantly. In addition, the use of certain materials for packaging has become much more efficient -- for example, the amount of aluminum it takes to make a food or beverage can has been drastically reduced since the 1960's -- and reuse and recycling of certain materials has become routine in many parts of the country.

What has happened in the past year or so is that the public, and various groups that serve the public's interest -- government regulators, consumer activists, and the business community -- have, once again, become aware that there is still a need to address solid waste issues. But just as the problem did not spring up, crisis-like, overnight, we cannot expect to solve it overnight. Certainly, we should not respond through our legal and regulatory processes with hasty, ill-conceived "solutions" that create additional problems, even if unintended.

Also to set the stage, let me turn for a moment to a description of the important role of packaging for food.

For some consumer products some packaging might be described as discretionary. By and large, however, packaging for food serves critical functions relating to product safety and quality. First, and unquestionably, neither legally-mandated nor voluntarily-undertaken changes in the way food is packaged can compromise the public's safety. Second, to the extent that food packaging enhances product quality and reduces spoilage, it contributes to a reduction in solid waste, since spoiled food becomes part of the municipal solid waste stream. It would make no sense to ban certain types of packaging, which represent the most effective means of preserving food, when the result would be to increase food waste. Some regulatory proposals would do just that. Similarly, it makes no sense to criticize or penalize small or "single serve" containers at a time when, demographically, consumers who want and need those types of food containers are increasing in the U.S.

The U.S. food supply is the safest in the world, and there is far less food waste here than in most, if not all, other parts of the world. We should not hastily jump at proposed solutions that only shift the type of waste we create. It is essential that as we all strive to deal with the solid waste problem, we recognize the special issues involved with food packaging.

Now let me turn to current regulatory activity and explain how GMA is responding.

You may know that some states have already enacted laws or regulations banning certain types of packaging and imposing strict controls on the use of environmental marketing claims. Moreover, on an almost daily basis, new bills are being introduced in state legislatures on these same and related solid waste topics.

Following are just some of the recently-enacted and proposed state laws:

o   In Maine, aseptic packaging, used primarily for juice drink "boxes", has been banned.

o   In Oregon last year, an initiative was on the ballot to require all product packaging to be "enviromentally sound" by January 1, 1993. The requirements of the proposed law would have resulted in de facto bans on a number of packaging materials. The initiative was defeated, but similar proposals are under consideration in Massachusetts and New Mexico.

o   In Iowa, there is a proposal to create a Packaging Review Board with the power to ban certain types of packaging.

o   California passed a law which establishes restrictive
    definitions for the use in advertising of words such
    as "recyclable", "recycled", and "biodegradable". The
    word "recyclable" is defined to mean that an article
    "can be conveniently recycled ... in every county in
    California with a population over 300,000 persons."

o   New York and Rhode Island have enacted laws which
    define environmental terms in the context of recycling
    "emblems". The New York definition for "recycled"
    products differs significantly from California's,
    thereby making it essentially impossible for a
    national manufacturer to distribute product bearing
    "recycled" terminology on its labeling in both states
    using the same label.

Many other public and private bodies are working on similar
proposals. This explosion of state and local activity causes an
enormous problem for national manufacturers. Through our state affairs
staff, GMA works actively to oppose efforts, such as the Oregon
initiative, that propose to address real solid waste problems by
unrealistic means. With respect to the regulation of environmental
marketing claims, we are working, along with a large group of other
trade associations and individual companies, to urge the Federal Trade
Commission (FTC) to promulgate environmental marketing guides, which
would apply uniformly throughout the country.

I would like to describe in more detail the regulatory events that have taken place and are developing in the environmental claims area. One significant regulatory development was the issuance of a document last November, called "The Green Report", by a task force of ten state attorneys general. This report proposes the establishment of a set of guidelines for the use of environmental marketing claims in the advertising and labeling of consumer products. The report calls for uniform, national guides and federal government leadership. It specifically calls on the FTC to assume the lead role in the regulatory process. This is such a significant development because the task force responsible for the report consists of essentially the same group of state AG's who have been acting independently of the federal government during the past decade to bring enforcement actions against national advertisers, for nationally-distributed consumer goods. So, the call for federal action and uniformity is a positive sign, from the perspective of businesses that market their products nationally.

In December of last year, the task force held a hearing to elicit public comment on its report. GMA testified at the hearing and made the following points. First, we wholeheartedly supported the primary recommendation of the report -- that the FTC, working with the Environmental Protection Agency (EPA), promulgate national guidelines covering environmental marketing claims for consumer products. We also concurred in the task force's recommendation that the states could have a role to play in formulating such guidelines. We agreed with the task force's goal of encouraging truthful and non-misleading practices, and its recognition that the proposed federal regulatory guidelines should be consistent with existing laws governing false advertising and deceptive trade practices. Clearly, the foundation for guidelines on environmental claims should be the deception and advertising substantiation policies of the FTC -- which are the law of the land with respect to advertising regulation.

While we agreed with many aspects of The Green Report, GMA took exception to the recommendation that "disposability" claims -- such as the term "recyclable" -- should be prohibited unless the "advertised disposal option is currently available . . . in the area in which the product is sold". This recommendation is apparently based on the task force's conclusion that "advertising today of an environmental benefit that cannot be realized until some uncertain time in the future is confusing and misleading to the public". We pointed out that even if there were some evidence (and the report cited none) that consumers commonly misinterpret disposability claims, the cure is not prohibition, but rather disclosure of adequate information to correct any misperception. Otherwise, any educational value or motivation to recycle, inherent in the "disposability" claim might be lost. We also noted that this proposed requirement to have disposability options (such as recycling centers) currently available in every community in which a product is marketed is directly contrary to the task force's support for national standards. Under the terms of this recommendation it would be impossible for companies to market products with disposability claims on a national basis. Economic and distribution considerations would simply preclude separate advertising or packaging for different geographic areas, resulting in a de facto ban on disposability claims.

GMA has followed up in the following ways on its announcement at the AG hearing to vigorously advocate the promulgation of FTC guides. First, we participated in the work of a coalition of approximately 20 trade associations to draft a petition to the FTC. The coalition consists of associations that represent the food manufacturing industry, the advertising industry, non-food manufacturers, retailers, and packaging materials manufacturers. The petition includes the text of proposed industry guides which we are asking the Commission to promulgate. The proposed guides cover topics such as claims of recycled content, recyclability, compostability, source reductions, reusability, and general claims. They do not attempt to provide definitive answers to scientific questions, but rather are intended to give useful guidance to the industry, through the use of examples, on how the FTC would be expected to respond to certain types of claims.

In addition to working on the petition itself, GMA also wrote to and visited with the FTC Chairman and Commissioners to impress upon them why we believe guides are needed and why quick action is desirable. We told them that absent Commission leadership in this important area, we foresee a continuation of local government regulation that may be inconsistent at best, and irreconcilable at worst. We pointed out that such inconsistency could lead to one or more of the following outcomes: (1) consumer confusion and skepticism about the meaning and value of claimed environmental benefits; (2) reluctance on the part of manufacturers to make environmental claims because of divergent local standards for nationally marketed products; and (3) diminishment of the incentive businesses would otherwise have to develop new products with environmental benefits, because they have been deprived of the opportunity to inform consumers of those benefits. We told the Commission that we expect FTC guides to become a point of reference for state and local enforcement, thus reducing, if not eliminating, the problems of inconsistent regulation.

I would like to leave you with a few concluding thoughts on the two regulatory issues that are most important to GMA -- the need for uniform, national standards, and the need to recognize the value to the consumer of truthful information disseminated through advertising and labeling.

The vast majority of consumer products are marketed in this country on a national, rather than local or statewide, basis. Therefore, it is especially important that rules for the use of packaging materials and the dissemination of information, through advertising and labeling, about the environmental attributes of products and packaging be uniformly applicable throughout the country. A single set of rules will benefit not only the industries that market products across state lines, but -- most importantly -- consumers. There is no reason to believe that consumers in one state understand environmental claims differently from consumers in another state. In addition, consumers as well as products travel freely throughout the U.S. So consumers will necessarily be confused if the same words are subject to different requirements in different states.

The solid waste problems for which we are all seeking solutions did not occur overnight and they were not caused by a single sector of our society. The Green Report states that "the environmental problems facing the world today are largely the result of the way we do business". I would add to that statement, that the way we do business in the U.S. is largely driven by consumer demand. Business really does exist to serve the consumer's interests. Companies do not and could not package or sell products that consumers do not need or want. Increased consumer sensitivity to the environmental implications of their purchasing decisions is a positive development. It is equally important to recognize, however, that the desire of industry members to inform consumers about the environmental features of products is also a positive development. Enhanced environmental awareness on the part of consumers will be of little value if they are unable to obtain the information they need to make purchasing choices based on environmental considerations. Markets work best when sellers are free to communicate the benefits of their products honestly and truthfully. At this early stage in the development of environmental marketing, overly stringent regulation, such as that passed or proposed in some of our states, will do far more than limit deceptive claims. It will dampen, and perhaps eliminate, incentives businesses otherwise might have to invest in the development of new packaging technologies and other environmental improvements. These facts should be the foundation on which any regulation of environmental claims is built.

The food and packaging industries can and certainly will play an important role in helping resolve solid waste problems, but a number of changes will have to occur over time -- including technological advances, consumer education, and government action. There is no single or simple solution to the country's solid waste problems.

SHOULD WE OR SHOULDN'T WE? THAT IS THE QUESTION

Carol D. Scroggins
President
The Consumer Voice, Inc.
3441 W. Memorial Road, Suite 5
Oklahoma, OK 73134

Biography

Carol Scroggins is President of The Consumer Voice, Inc., a research and marketing company focusing on the identification and management of consumer and quality issues. Scroggins formerly served as Vice President of Consumer Services for Fleming Companies, Inc., a wholesale food distribution firm serving more than 5,200 supermarkets in 37 states.

Ms. Scroggins has been involved in consumer and quality assurance relations for almost 20 years. A sought-after speaker, she has appeared before a number of prestigious educational and food marketing organizations, and provides consumer input and marketing expertise for trade associations, industry groups, food manufacturers, wholesalers and retail organizations.

Ms. Scroggins is a past chairman of the Consumer Affairs Council of the Food Marketing Institute (FMI), a non profit trade association of food retailers and wholesalers. In addition, she is active in numerous professional organizations, including the Society of Consumer Affairs Professionals in Business, the Society of Nutrition Education, American Home Economic Association, Home Economists in Business, International Platform

Association, American Institute of Wine and Food, Product Marketing Association Consumer Council.

She has served on the Boards of Trustees of the Missouri Home Economic Education Foundation, Ballet Oklahoma, Food Research and Action Council, and as a member of the Alulmni Board of the College of Human Ecology, Kansas State University. She currently is a board member of the Oklahoma City Art Museum and a member of 1990 Leadership America, sponsored by the Foundation for Women's Resources. She is a member of the Agribusiness Promotion Council (APC), a Private Sector Advisory Committee to Secretary of Agriculture, USDA. She is featured in the 1989-90 editions of Who's Who of Women Executives and Who's Who in America's Outstanding Women of the Eighties, and the Who's Who Registry.

A native of Texas, Carol Scroggins earned a BA degree in home economic education from Washburn University and an MS degree in family economics from Kansas State University, and has done postgraduate work at the University of Kansas.

Abstract

Consumers today are more demanding, knowledgeable, sophisticated, price sensitive, and confused!! They have become quite adept at finding where/how to get what they want. Should we or shouldn't we be focusing on consumer acceptance as one of the facets in responding to the packaging and labeling issues.

Supermarkets and private brand companies must play a more proactive role in label and package issues that affects customer attitude.

"SHOULD WE OR SHOULDN'T WE? THAT IS THE QUESTION"

Consumers today are more demanding, knowledgeable, sophisticated, price sensitive and confused! They know what they want and are willing to pay for it. And, they have become quite adept at finding where/how to get what they want. Unfortunately, they also have a growing cynicism because consumers <u>expect</u> that products and services (clothes, cars, televisions, household equipment <u>and</u> food) will be less than they want and less than what they got in the past. In short, today's shoppers generally believe they won't get the level of service or quality they want in product or service.

Quality has been identified by almost every futurist and business writer over the past five years as growing in importance. When looking toward the next decade, quality is proposed as a way to separate the good from the not-so-good. It raises the question, "Should we or shouldn't we?" Can we or can we not do more to identify and meet consumer expectations?

If quality is defined as what consumers want and are demanding in products, packaging, labeling or personal services; service is putting those needs and demands first. Therefore, quality service is identifying what consumers want, are willing to pay for and putting these issues at the top in decision making. The good news is that almost any sincere effort to provide quality in product and service will make a difference. But do not try to "snow" today's shoppers. It will be at your own peril.

If the expectations are not met, the customer simply may not buy the product again or be seen again by the store. According to Food Marketing Institute (FMI), approximately 30% of shoppers change stores each year. Statistically, this points out the life cycle of a shopper is 3+ years. This is more critical for private brand (PB) than national brands (NB) since PB is more limited in scope and availability. Gaining a new customer is five times as expensive as keeping a satisfied one, and one way to keep those customers is to listen to what they want from a shopping experience.

Private brand has traditionally not been seen as creative, innovative risk takers when it comes to meeting customers perception of quality in product standards or labeling. At some major companies, however, this has been changing. Since I have spent the better part of the past twenty years at a major wholesaler, many of my examples will be drawn from that experience, and from other clients. However, while functioning as a special advisor to several supermarket clients, I do not, and cannot, speak for them. My role here today is to discuss how supermarkets, including their brands, should respond to the packaging/labeling changes anticipated in the next few years.

Business management consultant, H. T. Garvey once said:

"When a business firm attempts to mold its whole policy to meet the prices of its competition that business is entering a labyrinth, the center of which is the chamber of despair. Highest quality never can be given nor obtained at the lowest price. If price is sacrificed, quality will also be sacrificed. If quality is sacrificed, society is not truly served."

Past history of the PB industry has held the prevailing attitude that seems to have been: "Errors are to be expected." They have often been accepted as a part of normal business activities. That means the focus has not been on meeting consumer's needs and demands. Should we or shouldn't we have the same goals for private brand that are expected of national brands? Obviously, to a private brander and his customers, the answer should be yes.

Identifying and providing consumers with the basics that they expect and demand from the food marketplace is really more of an ART than a science:

A - Accurate information about products and their uses

R - Responsiveness to concerns

T - Timeliness, customer time not necessarily business time

The goal of a consumer-focused program must include a measurement of excellence. The quality mission for PB should include a partnership between product development, label design and information, production, procurement, marketing, consumer, sales, delivery and evaluation. This means that attention must be focused on what the label "says" as well as "how it looks". Perhaps the "A" in ART is of Customer Focus.

PACKAGE AND LABEL INFORMATION

PB labels should be designed and implemented with the same care as any other. That means policies should include:

- graphics and design

- required information
    - legal and "voluntary"
    - evaluation and satisfaction
    - proactive or reactive

GRAPHICS AND DESIGN. There should be no different standards for PB than those expected from NB. Each company must establish rules consistent with company policy. It is then the responsibility of the designer to implement those rules and policies. Unfortunately, designers sometimes get caught up in the design and forgets the user's needs. Both can and should be meet in the presentation of the package or label.

PB graphic designers consider that established "look" of an item in designing the new. It is a delicate balance to bring new into conformity with existing in a creative way without coping. Generally, they have done a good job. A strong consumer voice can assist designers in what optional information should go on the label and how that part of the label should look.

A recent study reported by PRIVATE LABEL MAGAZINE, EXECUTIVE EDITION, says that:

> "75% of consumers feel the 'quality' of ingredients is the same between national and private brands. 50% say the information label is comparable. But some consumers see private brands as lacking variety and long term consistence. No longer is price the only factor in product selection."

Consumers make buying decisions based on observations and experiences, as well as, personal references. "Mother did or did not", often makes a great deal of difference in whether a shopper is willing to try new brands or products. The graphic design of the label helps to determine customer perception of the quality.

INFORMATION. Legal information required on labels is mandated by government agencies, whether local, state or federal. It is of vital importance to have one uniform set of label rules. Other kinds of label information is mandated by consumer expectation. The dilemma faced is getting these two requirements into some kind of acceptable accommodation.

When the first voluntary nutritional labeling regulations were drafted in the early 70's, many PB companies took "a wait and see position". Others who had established a consumer focus, like Fleming, were aggressive in making it a part of the company position. At that time, the biggest obstacle to overcome was the reluctance of suppliers to provide the information.

I wish I could say that has changed over the years. . . but it hasn't. There are still some NB and PB processors who are not eagerly embracing or preparing for changes that should be made. I believe that one of the reasons that we now face mandatory nutrition labeling is the lack of information. Another problem was the wide proliferation of health claims that occurred in the late 80's.

Much of the first additional information to Fleming's PB was due to consumer concerns and input. For example, sodium was not a part of the first voluntary nutrition labeling regulations. It wasn't long however, before the consumer department began to get the letters and phone calls requesting that sodium be added. Many were from senior citizens who had diet restrictions.

In the mid-seventies, sodium was sometimes used to enhance taste, or camouflage the lack of it. Since FDA has established regulations concerning sodium, and the trend toward changing taste preferences and health concerns, sodium levels have changed. I think that many companies were shocked, or at least a bit chagrinned, to learn just how must sodium was in their product.

The new mandated nutrition labeling, while it is seen as a hardship by some small companies, will make it easier to get all processors and manufacturers to do the right thing. It is the law and will soon have regulations and format established. My concern is that the same information on processed and packaged goods will be forced to be put on fresh products. There is no disagreement that FDA should establish rules for voluntary compliance in produce and seafood, but I hope that the rules have some flexibility to provide the information consumers need in these two new areas. Is it the same as processed? I do not believe that it is.

The new regulations will make it easier for a PB companies to respond with label information for their buying public. Regulation is still most likely 2½ years away. That seems plenty of time to get ready. But, I fear that time may be used up waiting and then a cry will go up about how much it will cost to get the information ready to put on labels because the analysis is not completed or the label/package inventory is too large.

Our challenge is to begin the preparation now in order to be ready when the final regulations are in place. Since there is still a lot of uncertainty, companies must start to plan now to be ready. It also means designs revisions underway or planned for the next year should keep the likely new rules in mind. Plan for a way changes can be made on as few color plates as possible to minimize cost.

USDA. Much of the discussion about changes has been focused on FDA and products that are governed under their "charter". Still another part of labeling to consider is the lack of uniformity between FDA and USDA regulations. For companies, especially PB, this means having to know and keep the two sets of rules separate. Many PBranders rely of the processors or packers for this type of help. While their help may make the label approval process at USDA easier, it is still difficult.

PB retailers and wholesalers are not excused by their customers. Companies still have the responsibility for what goes out with their name. Those companies who process/package meat items covered under USDA regulations should start right now to develop the nutrition information that is likely to be required either by regulation or consumer demand.

The industry should also be prepared for the possibility for "voluntary" guidelines in fresh meat/poultry items similar to the current "voluntary" guidelines in produce and seafood. Can this extension be far behind? Should it be?

FTC/EPA. The Environmental Protection Agency (EPA) records show that containers and packaging waste amounted to 56.8 million tons in 1988. This is the largest portion of municipal waste generated annually. It is also the most visible, much of it lands on our streets and parks. Consumers perceive that packaging is a big problem, therefore it is. So far, EPA has not developed guidelines or definitions for environmental claims.

Federal Trade Commission (FTC) does seems to be getting into the act on labeling by establishing a set of rules to govern what can be said about solid waste/environmental claims. A uniform standard set of rules would be welcomed. But it will continue to be more difficult to design labels that can make room for all the information that must be made and extras that consumers would like to have.

Several states have and are passing legislation that is viewed by many as "shooting from the hip" in order to do something. One set of environmental rules would be far preferred to different ones by state, local or a multi-governing unit regulations. Companies that operate in more that one state will be especially hard pressed to do the right thing.

Technology is changing so fast that what is acceptable today, is not acceptable tomorrow. In eagerness to do something, that is sometimes seem as self serving and misleading. Environmental labeling is very difficult.

So far FDA has not allowed recycled plastics to be used in food containers except foam egg cartons. But if Pepsi and Coke have their way, that ruling may be a thing of the past. Because those two companies recently announced, separately of course, that they were planning to use bottles made from recycled plastic. One of the biggest battles may be securing the recycled plastic. Consumers are not yet in the groove of putting two or three liter bottles in the recycling bins and there not enough recycling facilities available.

If the used P-E-T bottles are recycled at the grocery store, it may be as big a mess as the recycling bins on the parking lot for newspaper, have been in some locations. I was recently by a supermarket parking lot that accepted newspapers for recycling. They were trying to be corporate citizens. The parking lot looked like the dump. There were signs everywhere saying, "NEWSPAPERS ONLY, PLEASE!!!", but it looked like a local garbage truck had a case of upset stomach. This particular store didn't give up. But they did have to assign someone to police the parking lot several times a day and night.

The food label of tomorrow may look like to one that was suggested in the 70's. This tongue-in-cheek approach was one that wrapped around, and around, and around a #2½ can . . . you do remember the #2½ can? We might see something like this, for real, if all the information required, suggested or hinted at, are made a part of the food label.

SUMMARY and CONCLUSIONS

Consumers are more sophisticated and knowledgeable that ever. Changing demographics, and lifestyles, may give us a future of even more knowledgeable, and often intense, customers. Today's shoppers have developed a cynical attitude that has lowered their expectations about the quality of service or products that they want. What will be our recourse with the shoppers of tomorrow if we cannot/do not respond to the ones we have today?

Consumer expectations are so low that the good news is that exceeding those expectations becomes relatively easy. Sincere efforts to provide quality products/information and soliciting consumer's input can make a big difference in their acceptance and loyalty.

Identifying and providing consumers with the basics that they demand from the marketplace is more of an ART than a science. Accuracy, responsiveness, and timeliness are key ingredients in a consumer-focused company, whether that company is private or national brand orientated. How the food industry responds to this ART with accurate, timely information on the new labeling requirements may be seen as a reflection of a commitment to quality in services and products.

Issues that will require answers include nutrition, diet and health, environmental products and services offered at retail. Should we or shouldn't we actively listen to shopper desires, demands in food packaging and labeling? I say yes!!

"FOOD MANUFACTURER/PACKAGING SUPPLIER:
SHARING THE BURDEN OF PROPOSITION 65"

Luther C. McKinney
Senior Vice President - Law
Corporate Affairs and Corporate Secretary
Quaker Oats Company
321 North Clarke St. Suite 2710
Chicago, IL 60610

Biography

After receiving an undergraduate degree in Agricultural Economics from Iowa State University and a law degree from the University of Illinois, Mr. McKinney left a senior partnership to join Quaker in 1973. He is responsible for the areas of law, investor relations, government affairs and corporate communications and is a member of Quaker's Board of Directors and Executive Committee.

Abstract

California Proposition 65 necessitates a symbiotic relationship between food manufacturer and packaging supplier analogous to a food and its container. Uncertainties about exemptions or safe harbors and the primacy of federal or state safety requirements for foods and their packaging, expose manufacturers and suppliers alike to the risk of litigation, civil penalties and adverse publicity. To avoid a warning scenario, suppliers should review listed chemicals, do risk evaluation or analysis work and share information with the food manufacturer. To the consumer, packaging is inseparable from the food and its brand or trademark. Our industries are therefore exposed to somewhat different levels of risk; burdens best shared by exchange of appropriate information and mutual cooperation.

Thank you Jerry (Heckman). It's a pleasure to join you and the other panelists to discuss regulations for plastics packaging and what's to come.

My remarks will reflect a food manufacturer's view of the burdens we share with packaging suppliers under California Proposition 65, an initiative statue properly known as "The Safe Drinking Water and Toxic Enforcement Act of 1986." The implications of the scientifically unfounded toxic warning requirements of Proposition 65 necessitate an almost symbiotic relationship between the food manufacturer and the packaging supplier that is analogous to a food and its container.

Proposition 65 was designed by its authors and supporters to limit the California public's exposure to known carcinogens and reproductive toxins from drinking water, consumer products and other sources. The law requires California to publish a list of chemicals known to the State to cause cancer or reproductive toxicity. Discharge of such chemicals into drinking water or a source thereof is prohibited. Of greater importance to our industries, no person in the course of doing business may knowingly expose any individual to a listed chemical without first providing a clear and reasonable warning to that individual.

There are three statutory exemptions from the warning requirement. Federal preemption is one, but more on that later. Another is the 12 month grace period after a chemical goes onto the California list. The third exemption applies if the exposure poses "no significant risk" assuming lifetime exposure to a carcinogen at a specific level, or if the exposure has "no observable effect" assuming exposure to a reproductive toxin at 1,000 times the no observable effect level. By regulation, the State defined "no significant risk" using a one in 100,000 risk formulation.

Proposition 65 implementing regulations contain an additional exemption from the cancer warning requirement for products regulated by the Food and Drug Administration that meet federal and state safety requirements. However, that exemption was legally challenged by Proposition 65 proponents and the trial court has found this exemption to be invalid. The court's decision is stayed during review in the California Court of Appeals. An appellate court decision is anticipated sometime in mid-year. In any event, the regulatory exemption from a warning on FDA-regulated products does not apply to reproductive toxicants.

FDA regulates the safety of food packaging under the concept that complex evaluation may permit the clearance of products after reviewing their safety under intended conditions of use. That is, you identify the intended use of the packaging for a type of food under certain conditions, temperatures, and so forth. Extraction studies determine whether there is migration. If so, study results are extrapolated to arrive at an estimated dietary intake level for the migratory component. The calculation requires sophisticated scientific work by FDA, a capability unique to the agency. Under current and challenged Proposition 65 regulations, packaging that survives FDA scrutiny is presumptively exempt from cancer warnings if; 1) there is no migration, 2) the material is generally recognized as safe (GRAS), 3) the material or component is the subject of a prior sanction or approval, or 4) is the subject of an applicable food additive petition.

If the cancer warning exemption for foods and food packaging is sustained in the appeals process, food manufacturers and packaging suppliers are spared some of the

Proposition 65 burden. But the relief is temporary. State regulations provide that the FDA exemption will phase out as specific "no significant risk" levels are established by regulation for some 50 priority carcinogens. A number of such "safe use levels" have been promulgated and others are in process.

The "safe" food in a "safe" container remains at unknown risk to Proposition 65 compliance actions because of the law's reversal of the traditional burden of proof from the plaintiff to the defendant. Compliance actions may be initiated by environmental groups seeking selective enforcement or the awarding of civil penalty "seed money" with which to prepare for future compliance litigation. The largest fine in the first four years of the Act, $750,000 against retailers who failed to warn for non-cigarette tobacco products, produced an award of $75,000 for Proposition 65 proponent groups. This case demonstrates that Proposition 65 puts a food and its container potentially at risk for not only the manufacturer and the packaging supplier, but for retailers as well.

We'd like to see our packaging suppliers be a part of supply chain management when it comes to Proposition 65 compliance. Supply chain management is the process of integrating the operations aspects of the business, such as purchasing, manufacturing and distribution, with the "go to market" practices, such as ad campaigns, trade deals and consumer promotions, in order to achieve the best outcomes. These outcomes include the lowest possible delivered costs to the consumer and the best possible supplier and customer relationships. Cooperative and mutually beneficial work by food manufacturers and suppliers on Proposition 65 problems is a good example of how supply chain management can benefit all parties, including consumers in California and elsewhere.

I stress the word cooperation because I hope that our industries can avoid the sort of "materials war" among trade groups representing various packaging mediums on issues of environmental and ecological concern. Proposition 65 is too different from routine environmental protection legislation and regulation and too risky for both our industries to allow mistrust to creep into the food manufacturer and packaging supplier relationship.

Proposition 65 forces the food manufacturer to evaluate the legal and public relations exposure of not only our products and their brands and trademarks, but our packaging as well. If migration is a problem for us, you share the problem. If the package itself creates an exposure, you have a problem, but we share it. From the retailer's point of view, we both have a problem in either case. The retailer might simply want our product off the shelf or may demand an on-label warning. Additionally, and of significance to both the food manufacturer and the retailer, the consumer doesn't differentiate between the food package and the product and its brand or trademark. A warning scenario because of food packaging material is likely to be perceived by the consumer as a food product warning.

The question naturally arises, how do we avoid the warning scenario in California and elsewhere? (Proposition 65-like bills have been introduced this year in both New York and Illinois, and can be expected to surface in Massachusetts.) First and foremost, we have to be familiar with the statute and its implementing regulations, court cases and FDA's responses to Proposition 65 problems. That means we have to keep up to date with how Governor Pete Wilson handles Proposition 65 implementation.

The Wilson Administration gives environmental concerns more priority than former Governor Deukmejian did. This focus began during the campaign when Pete Wilson refused to support either Big Green or a pesticides counter-initiative and promised to establish a statewide California Environmental Protection Agency. Gaining approval for CALEPA will not be without controversy, especially since Big Green's supporters are going to the legislature with the same menu of environmental nostrums that were in the initiative. Proposition 65 supporters will also keep the pressure on the Wilson Administration to implement the statute in ways favorable to their point of view.

How the implementation of Proposition 65 fits into the Wilson plan and which people will run the program are very important issues for food manufacturers and packaging suppliers. Our industries need to stay on top of present and proposed regulations, the affairs of the Proposition 65 Scientific Advisory Panel and the State's role as defendant in a number of significant law suits.

The California's Attorney General Office also changed hands in January. Former Attorney General John Van de Kamp was strong on environmental litigation and worked closely with Proposition 65 proponents on compliance suits. Van de Kamp refused to represent the State when environmentalists and labor groups first challenged Proposition 65 implementation by attacking the Governor for initially listing only 29 chemicals as known to cause cancer or reproductive toxicity. The Governor lost that suit, but the listing process has resulted in approximately 400 entries thus far, mostly carcinogens. The new Attorney General is expected to be more active in supporting the Governor's positions on Proposition 65 issues.

Attorney General, Dan Lungren, appears to be a classic conservative who enjoys good relations with many of the county district attorneys. Unlike Van de Kamp, Lungren can be expected to help protect the Governor with both advice and legal defense. Lungren appears unlikely to collaborate extensively with activist environmentalists. On the other hand, if the AG fails entirely to prosecute Proposition 65 enforcement cases, the California trial bar and activist groups will probably step up the number of 60-day notices for compliance suits in hopes that the AG and county or municipal prosecutors will leave the field of compliance litigation open to private prosecution and bounty hunters.

Food manufacturers and packaging suppliers must regularly review the list of chemicals known to California to cause cancer or reproductive toxicity. In fact, we need to review the chemicals which are to be considered for listing. If there are good scientific or policy reasons for making representations to the State or the Proposition 65 Scientific Advisory Panel on a particular chemical, we need to assure that representation is promptly and effectively undertaken.

We also need to examine each of the listed chemicals to screen for those which are or could be present in a food or in food packaging. Looking at doses, routes of exposure, duration of exposure and the like, we need to determine which listed chemicals are of interest to food products. Chemicals that might be or are present in foods must be evaluated in light of the implementing regulations for Proposition 65 and in relationship to the safety evaluations of the Food and Drug Administration.

For this reason, a strong cooperative relationship between the food manufacturer and packaging supplier is essential. In my view, that relationship relies on a willingness to share significant information. For example, if a component of food packaging appears on the Proposition 65 list, a food manufacturer will need more information. This could include information used to obtain FDA safety clearance or food additive status for the component, the intended conditions of use, estimated dietary intake levels and the like. In some cases, the food manufacturer may have some of this information because of previous cooperation in the preparation of FDA submissions on packaging materials.

The packaging supplier and the food manufacturer must evaluate whether there has been any essential change in the intended conditions of use with regard to specific food-product applications for the packaging material, component or system. This may help answer the question of whether there is an exposure under Proposition 65 or how it might be avoided.

An essential question is whether a packaging component migrates into the food, thereby making the food itself the agent of exposure. When it comes to the safety, reputation, and good will of a branded food product, the manufacturer is going to want to find a way to protect its investment in those values and their worth in the national market, not just sales attributable to California. He needs help from the supplier in answering the migration question and in finding ways to avoid this problem.

For the food manufacturer, I think it would be difficult to justify putting the food product at risk by accepting a warning which pertains only to its package. In the case of exposure through migration, we'd certainly look for maximum levels of technical assistance to find a way to retain the supply relationship while at the same time avoiding a warning scenario. Such a relationship exists today in the evaluation of microwave susceptor board.

If current regulatory or statutory exemptions or safe harbors have been applied and all packaging or product modifications designed to reduce or eliminate a listed chemical have been accomplished, continued cooperation would be necessary to determine whether the remaining level of a listed chemical meets the no significant risk or no observable effect standard. Risk assessment data used in obtaining FDA safety clearance for packaging could be useful in deriving a specific "no significant risk" or "observable effect" number. Such risk level information, if consistent with principles elucidated in the statute and implementing regulations, can be used as a defense in compliance suits. It also could be helpful in searching for ways to eliminate or reduce the level of a chemical in the food or food package.

Those are some of the specifics, but here are some general ways in which food manufacturers and package suppliers can cooperate in order to share the burden of Proposition 65.

Proposition 65 shows how the tyranny of a few drafters and the operation of the initiative process can be used to further the popular belief that no harm is too small or to inconsequential to warrant total mobilization of the State's and industry's resources to "cure the problem" of toxics. We need to work on ways to rationalize the notion that risk comes in one size and one size only: extra large. For example, because of industry advocacy, the reproductive toxicity risk presented by tiny amounts of vitamin A, an essential human nutrient, resulted in the listing by California of this chemical only if the level exceeds twice the U.S. Recommended Daily Allowance. Otherwise, the 1,000 fold safety factor would have resulted in reproductive toxicity and birth defect warnings on common foods like milk and fortified cereals. Warnings like that would have resulted in vitamin A intakes among women far below current public health recommendations.

We need to support public education efforts in the areas of comparative risk, science and technology and special training for those working in the fields of risk evaluation, assessment and communication. We need to provide honest, correct information to the public when real or imagined foods safety incidents arise. But these are long-range steps. Other more concrete actions are desirable too.

Food manufacturers have undertaken, with the help of Jerry Heckman and the Society of the Plastics Industry, a law suit in the San Francisco Federal District Court challenging Proposition 65 food warnings and seeking broad preemptive relief. The suit, brought by the Committee for Uniform Regulation and Labeling (CURL), is in discovery, but the case is going forward. One of the drafters of Proposition 65 once described the CURL suit as a hand grenade waiting to go off. I'm glad that Jerry and SPI are assisting in this important litigation because it might be the last hope for preemption.

On the legislative side, Congress last year refused on germaneness grounds to include coverage of Proposition 65 warnings in the preemption section of the Nutrition Labeling and Education Act. Despite arguments from our side that the nutrition labeling bill was all about diet and disease, the House Energy and Commerce Committee Chairman decided that the issue of toxic warnings on foods was really a food safety problem not suitable for a nutrition bill.

While we didn't get a vote on toxic warning preemption last year, I'm persuaded that Proposition 65 and the interstate and international trade burdens threatened by California's so-called "Big Green" initiative in California combined to help consolidate Congressional feeling that the misguided anti-toxin initiatives of a single state can have broad, pervasive and negative effects on markets and commerce at home and abroad. There's going to be some sort of food safety legislation in this Congress. With your support, we will be able to achieve express federal preemption for Proposition 65 warnings on food when an appropriately germane legislative vehicle is considered.

Should Proposition 65 implementation and litigation produce results that further damage food manufacturer and packaging supplier interests, and efforts to obtain preemption of toxic warnings by means of either congressional or judicial action falter, there remains another avenue of preemptive relief. That form of relief is what we call express administrative preemption; an executive order or a regulation which specifically strikes down state or political subdivision toxic warning requirements, or mandates a national, uniform federal standard having preemptive effect.

I'm encouraged that the climate for administrative preemption is improving. Big Green was defeated in California. Congress passed preemptive food labeling legislation. The Bush Administration supported uniform pesticide residue tolerances and advocated scientifically-based phytosanitary and sanitary international food safety standards. The European Community has opted for economic unification in 1992. These are all signs of recognition and support for harmonization of differing regulations and the need for uniform requirements. Even our friends in the European Community and in the National Association of Attorneys General (NAAG) Task Force on "Green Labeling" agree on the importance of uniformity. The European Community announced a labeling plan late last year, saying that it wants to unify its "green" labels from the start, rather than have conflicting labels later. And a NAAG task force last November recommended that the federal government set uniform industry standards for environmental marketing claims.

However, national uniformity is not just around the corner in the Bush Administration. Just last December the Justice Department filed a friend-of-the-court brief supporting a Wisconsin Public Intervenor and Town of Casey, Wisconsin, petition to the U.S. Supreme Court to grant certiorari on the question of whether the Federal Insecticide, Fungicide and Rodenticide Act (FIFRA) pre-empts local regulation of pesticide use. The Wisconsin Supreme Court upheld FIFRA preemption. The Supreme Court's decision in this case could be significant on express preemption involving state-required toxic warnings.

Our industries can help share and rationalize the burden of Proposition 65 and diminish the threat of purposeless diversification of state and local toxics warnings law. We need to support intra-and inter-industry communication and cooperation to facilitate a broad-based effort to get the Bush Administration to expressly preempt state and local government requirements for toxic warnings on FDA-regulated products. Our arguments are strengthened by the new leadership at the FDA and by Congressional willingness to assign the agency additional safety responsibilities each year, even though these responsibilities are not always accompanied by the level of additional financial support we'd like to see.

Of course, our preemption arguments would be strengthened if some unfortunate actually is forced to put a warning label on a FDA-regulated product that otherwise meets federal safety requirements. That would get the attention of not only the Bush Administration and Congress, but probably that of the trial bar and its dreams of product liability, class action and RICO litigation. If another state adopted similar but contrary toxic warning requirements we would also have a "smoking gun" that would compel federal action on preemption. Our industries must support the scientific, technical and experiential primacy of FDA as the World's leading product safety agency. To properly support FDA, we have to fight against Proposition 65 food warnings and their clones in other states.

Food manufacturers and its packaging suppliers have more than most industries to lose in the toxic warning game. Food manufacturers and their packaging suppliers do not have quite the same levels of risk under Proposition 65. But we know that a warning associated with either a packaged food or its package is bad news for both industries. By working together on Proposition 65 issues and advocacy, and sharing regulatory and product-specific information, food manufacturers and packaging suppliers are following good business principles and contributing to the solution of potentially serious food safety and consumer confidence problems.

Thank you very much.

FTC ACTIONS WITH RESPECT TO ENVIRONMENTAL CLAIMS

C. Lee Peeler
Associate Director
Division of Advertising Practices
Bureau of Consumer Protection
Federal Trade Commission
6th St. & Penn Ave. N.W.
Washington, DC 20580

Biography

C. Lee Peeler, Esq. has been Associate Director for the Division of Advertising Practices with the FTC since 1985. He joined the FTC as a staff attorney in 1973. During his career he has held a number of management positions, including Assistant Director of the Division of Credit Practices, Acting Associate Director for Credit Practices, and Executive Assistant to the Director of the Bureau of Consumer Protection.,

Abstract

Current activities and approaches to environmental claims in advertising and labeling.

REMARKS OF
C. LEE PEELER
ASSOCIATE DIRECTOR FOR
ADVERTISING PRACTICES
BUREAU OF CONSUMER PROTECTION
FEDERAL TRADE COMMISSION

BEFORE THE
PLASTICS INSTITUTE OF AMERICA, INC.
FOODPLAS CONFERENCE

MARCH 6, 1991
ORLANDO, FLORIDA

I appreciate the opportunity to meet with you today to discuss the Bureau of Consumer Protection's current activities and approaches to environmental claims in advertising and labeling, or so-called "green marketing."[1]

It is, of course, no secret that the environmental effects of products and their packaging have become increasingly important issues to American consumers. What is surprising is the apparent recent growth of that concern. In September of last year, Advertising Age reported a poll finding that 82% of those surveyed had changed their purchasing decisions based on concerns about the environment. In addition, 78% of those polled said they would pay up to 5% more for environmental packaging -- up from 64% from the year before -- and 47% said they would pay up to 15% more -- up from 27% last year. Similarly strong results are shown by other recent surveys. Moreover, the latest survey data shows that in addition to strong positive support by consumers for environmentally compatible products, there is also a strong negative reaction to companies whose products are perceived as environmentally harmful. Of those surveyed, 64% said they were less likely to buy from a company with a poor environmental record.

Given these strong consumer preferences, it is no wonder that marketers have been increasing their advertising and labeling of "environmentally friendly" products. This is the basic way markets respond to consumer demand and can provide

---

[1] The views I express today are my own, and do not necessarily reflect those of the Commission or any commissioner.

significant benefits. To the extent that advertising and labeling address environmental concerns, they will exert significant pressure on manufacturers and retailers to improve the environmental qualities of their products. However, to the extent that the advertising and labeling claims mislead consumers, the claims can lull consumers into taking actions such as spending money for products that may not help the environment or that, in the worst case, may injure it. Similarly, if consumers do make purchasing decisions based on environmental representations that they later learn are untrue or misleading, there is a risk that the credibility of environmental advertising and labeling, not to mention all advertising generally, will be undermined.

As you are not doubt aware, questions have been raised by a variety of groups concerning the accuracy of some claims that are currently being made. The Commission staff now has about 20 investigations underway to examine environmental advertising and labeling claims being made for a variety of products. I would like to discuss with you today some of the issues that these investigations have raised.

First, let me make clear that environmental claims in advertising and marketing are not a new phenomenon. The Commission has investigated and taken action against deceptive or unsubstantiated environmental claims many times in the past. For example, in 1973 the Commission filed a complaint and consent order challenging representations that a plastic coated paper milk carton would biodegrade when buried in a landfill.[2] A year later, the FTC litigated a major case against a gasoline manufacturer that made misleading fuel emission reduction claims.[3] And during the early 1980's, environmental quality claims for air filters and water filters were a staple of the Commission's advertising enforcement program.

A more recent example is the Commission's action against Vons Companies,[4] a major Southern California grocery store chain that sold produce with a "Pesticide Free" claim. Although Vons had made an extra effort to <u>reduce</u> the pesticide levels on the produce it sold, the produce was not "pesticide free."

Thus, environmental advertising and labeling have long been issues at the FTC. What is new, however, is the rapidly expanding consumer demand for "environmentally friendly" products that is indicated in the survey data I discussed earlier and is driving the marketing claims.

The FTC's interest in these claims will come as no surprise to those familiar with the Commission's approach to advertising regulation. Environmental benefit claims are precisely the kind of "credence" claims on which the Commission has traditionally focused, because consumers cannot easily evaluate them for them-

---

[2] <u>Ex-Cell-O Corporation</u>, 82 F.T.C. 36 (1973).

[3] <u>Standard Oil of California</u>, 85 F.T.C. 1401 (1974), aff'd <u>as modified sub nom. Standard Oil Co. v. FTC</u>, 577 F.2d 653 (9th Cir. 1978).

[4] <u>Vons Companies, Inc</u>., FTC Docket No. C-3302 (September 28, 1990).

selves. Advertised environmental benefits, for example, may not be evident for five, ten, fifty or more years, and even then may occur in ways that consumers cannot verify or measure. How can a consumer today judge whether a particular plastic garbage bag will actually degrade over time in a landfill, or whether a particular aerosol spray will harm the earth's ozone layer, or whether a paper towel really is made of 100% recycled fiber?

The FTC's interest reflects its statutory authority under the FTC Act. First, advertising and labeling must be truthful and not misleading. Second, objective claims about products must be substantiated by competent and reliable evidence. Both of these standard legal requirements apply with full force to claims that are expressly made and, equally important, to the reasonable implied claims that consumers take from the ad or label.

The focus of the FTC's current inquiries is specific product performance claims. For example, the staff is looking at whether products are actually "photodegradable," "biodegradable," "ozone friendly," "recyclable," or made from 100% "recycled" fiber. In addition to these specified claims, we are also reviewing more general claims such as that a product is "environmentally safe" or "environmentally friendly."

As in any other Commission advertising investigation, we are looking at two primary issues. First, what are the claims that are being made to consumers? Second, are those claims adequately substantiated? For example, when claims are made that a product is harmlessly "biodegradable," "photodegradable," or "degradable," we want to know whether the product will **break down**, under what conditions, how completely, and how fast. We also want to be sure that in the process of breaking down, the product does not itself produce harmful by-products.

We are also concerned that some environmental claims may be misleading because of common consumer misperceptions. For instance, consumers may not know that most of their garbage and trash is destined to be buried in sanitary landfills, or they may believe that landfills are biologically active environments in which garbage is composed. In fact, current landfills are managed to minimize biological activity, and there are well-known excavations of landfills which have shown that even such organically-based items as hot dogs, carrots, and newspapers may degrade very, very slowly, if at all, in many landfill environments. Thus, to the extent that consumers believe that degradable products will provide a significant solid waste reduction benefit, claims of degradability for products like plastic or paper trash bags and disposable diapers, which are ordinarily disposed of in landfills, may mislead the public.

We are also looking at claims made for aerosol products concerning the ozone layer. Claims such as "ozone friendly" or "CFC free" have the potential to be deceptive if, for example, the product merely has reduced levels of CFC's, or chlorofluorocarbons, or if it contains ozone-depleting chemicals other than CFC's. We are concerned about the potential for misleading claims in this area and we are working with the EPA to establish an enforcement presence.

As issues relating to degradability and ozone-friendly claims are resolved, other doubtlessly will be raised. Some of these will raise very difficult questions. For instance, in regard to "recyclable" labels, questions include the amount of recycling that needs to be in place near the point of purchase to support such a claim. Similarly, "recycled" labels raise questions as to the source and extent of the raw materials on which a "recycled" content claim is made.

In reviewing such claims, we are interested in what those terms mean to consumers. These issues, however, also implicate solid waste management policy concerns, and may not be appropriately resolved solely from an FTC deception standpoint. For example, it may be desirable to build in incentives to industry to use more recycled material by requiring that a product be composed of the highest possible percentage of recycled material before it can be called "recycled." This policy question, however, may be separate from that of whether consumers are deceived by the "recycled" label.

One noteworthy development in the green marketing field has been the advance of proposals for environmental labeling certification programs. Although such programs are one way consumers can receive information about the environmental benefits of products, we do not view certification programs as a simple answer to the complex issues of solid waste management and consumer deception. In addition, there are concerns that such programs themselves have the potential to be deceptive. Manufacturers of products that are part of the program must also ensure that seals are not used deceptively or anticompetitively. As such programs become operational, the Commission will review product advertising and labels to ensure that they do not result in deceptive or misleading advertising claims. Each such program will have to be evaluated individually for compliance with applicable federal law.

We are closely coordinating our activity in the environmental area with the Environmental Protection Agency, and we are relying on EPA's expertise to assist us in evaluating many of the technical scientific issues raised by environmental marketing. Our examination of degradability claims, for instance, has been guided in part by EPA's recently issued report on plastic waste management issues, which concluded that degradable plastics provide no real waste management benefits in terms of source reduction, recycling, landfilling or incineration. It is particularly important that the FTC closely coordinate with EPA in this inquiry because EPA has responsibility for many of the larger policy issues, such as overall solid waste management, that these inquiries raise.

The first priority for the FTC in this area is to eliminate deceptive claims in the marketplace. Our investigations aim to do just that. Our focus is the message that is being conveyed to consumers -- that new and improved products, or even old products, are providing significant environmental benefits.

Throughout this process we also have been working in close cooperation with a group of state attorneys general. The objective of this coordination is to achieve uniformity of results to the extent possible. The process has also been remarkable in that, from the beginning, the states have recognized that many of the issues raised are national advertising and marketing issues that would best be resolved on a national basis in consultation with the states.

In addition to the state attorneys generals' activities, we have seen significant interest in the issues by other state and local governmental units, including the Council of State Governments, and the Council of Northeast Governors. Many state legislatures also have entered the picture. As you are aware, legislation recently enacted in California establishes definitions for a variety of environmental marketing terms such as "biodegradable," "photodegradable," "recyclable" and "ozone friendly," as well as general terms such as "environmentally friendly" or "safe." And you know that making use of these terms in a manner inconsistent with the statutory definitions is a criminal violation in California.

As my remarks suggest, environmental labeling and marketing issues do not seem ready to disappear. It is important for marketers and advertisers utilizing such claims to be careful that both express and implied objectives product claims are sub-

stantiated. As the Commission takes enforcement actions on a
number of fronts, I believe that standards for these claims will
become clearly more predictable. We look forward to working with
you on a cooperative basis to promote a better understanding of
our policies in this area, and to ensure a marketplace where
environmentally concerned consumers can depend on green claims.

AN FDA PERSPECTIVE OF REGULATIONS FOR FOOD PACKAGING

Dr. Patricia S. Schwartz
Acting Director
Division of Food Chemistry and Technology
Food and Drug Administration
HFF 410, 200 C St. N.W.
Washington, DC 20204

Biography

* A.B. Degree in Chemistry from Bryn Mawr College, 1968
  Ph.D. Degree in Organic Chemistry from Yale University, 1974

* Worked for Monsanto Company 1974 - 1978 as a patent chemist

* Joined FDA in 1978. Currently the Deputy Director and Acting Director
  of the Division of Food Chemistry and Technology in the Center for Food
  Safety and Applied Nutrition

An FDA Perspective of Regulations for Food Packaging

Since 1958, the U.S. Food, Drug, and Cosmetic Act has mandated pre-market approval of all substances which meet the definition of food additive. A substance is a food additive if its intended use "results or may reasonably be expected to result, directly or indirectly, in its becoming a component or otherwise affecting the characteristics of any food." Components of food packaging materials can and do migrate into the packaged food. Consequently, food packaging materials are considered indirect food additives (to distinguish them from <u>direct</u> additives which have an intended technical effect in the food) and are subject to regulation.

The regulations for food additives are in the form of positive lists published in Title 21 of the U.S. Code of Federal Regulations. There are two very important principles inherent in these regulations that I want to emphasize today. The first principle is that the regulations are written to be generic, i.e., they stipulate the safe conditions of use of the additive, rather than licensing a particular company's product. This means that company B does not have to re-petition FDA <u>for the same use</u> of a food packaging material for which company A has already obtained a regulation.

The permitted conditions of use for the additive, which are stipulated in the regulation, may include temperature, types of food, use level in the plastic (for an adjuvant), and whether the food contact will be one-time or repeated (food processing equipment). Each of these parameters affects the level of migration and hence the potential human exposure to the additive. By stipulating the permitted use conditions, the regulation in fact is prescribing the level of exposure to the additive which has been determined to be safe.

Why is this important in terms of what's coming in food packaging regulations? In response to the American consumer's demand for convenience and speed in food preparation, the food package has taken on new roles. It now frequently serves as the container in which the food is cooked or reheated. With the development of microwave susceptor technology, the package can now participate in cooking the food.

These new roles mean that the food package is being subjected to higher temperatures. Since migration is proportional to temperature, these new technologies have the potential to result in higher levels of exposure to migrants from packaging materials than were prescribed by the existing regulations. The increased exposure resulting from the new use could mean that there is no longer an adequate margin of safety between the estimated daily intake (EDI) (the measure of exposure) for the additive and the acceptable daily intake (ADI) that was determined from the toxicological data submitted in support of the original regulation. The higher level of exposure could also mean that FDA would require additional toxicological studies to demonstrate the safety of the new use for the packaging material.

The key word here is demonstrate. It is not sufficient that the manufacturer be convinced that the new higher temperature use of a regulated packaging material is safe. If the new use results in significantly higher migration so that additional toxicological data are required to demonstrate safety, then the burden is on the manufacturer to provide those data to FDA prior to marketing the packaging material for the new higher temperature use.

Demonstrating safety for a new food packaging material or for a new use of regulated packaging material is the responsibility of the manufacturer. But FDA also has a responsibility in the face of the current advances in packaging technology. There are a number of regulations for food packaging materials in 21 CFR Parts 176-177 which have no limitations on maximum use temperature. At the time these regulations were written, it was apparently thought that the migration tests conducted by the petitioners had been carried out at sufficiently high temperatures to cover all possible uses of that particular packaging material.

This has proven not to be the case. One of the best illustrations of this kind of situation is the microwave susceptor. As I'm sure you all know, this innovative technology is designed to permit browning/crisping of a variety of foods in the microwave oven. It can also be used to increase the yield of popped kernels when popcorn is prepared in the microwave. Susceptors are manufactured in a variety of configurations (sleeves, trays, etc.) but generally the food contact layer is either poly (ethylene terephalate) (PET) or paper.

Neither the regulation for PET (21 CFR 177.1630) nor the regulation for paper (21 CFR 176.170) contains a specific maximum temperature limit. However, the data submitted to establish both of these regulations were based on applications involving significantly lower temperatures than those measured for the food contact layer in microwave susceptors. Studies done by FDA's Indirect Additives laboratory clearly showed that cracking and melting of the PET food contact layer occurred when susceptors were used to heat food according to package directions. The PET was not serving as a barrier to migration of components of the underlying adhesive layer into food.

In such a situation, FDA needs to clarify the regulations with respect to the permitted use conditions. With this goal in mind, FDA has requested from industry the kinds of data that are necessary to demonstrate the safety of the various types of microwave susceptors. FDA has received data from several individual companies as well as from the National Food Processors Association/Society of the Plastics Industry Susceptor Microwave Packaging Committee which was formed to fund research on migration from susceptors. Based on its evaluation of the submitted data, FDA will publish in the Federal Register an appropriate clarification of the permitted use conditions for the materials used in susceptors.

The microwave susceptor situation is by no means the only one requiring clarification of the permitted use conditions of regulated food packaging materials in response to advancing technology. Another example is the plasticized PVC cling films being used to cover food or food containers during microwave cooking. This is a rather complex situation because several "regulatory boxes" are involved.

Vinyl chloride homopolymer resins (meeting certain specifications on volatility and inherent viscosity) are prior sanctioned for use in plasticized films for food contact. This means that this use had been explicitly approved by FDA or USDA before September 6, 1958. Prior sanctioned uses of PVC are exempt from the food additive provisions of the Food, Drug, and Cosmetic Act and do not require pre-market approval. They are still subject to the adulteration provisions (Sec. 402) of the Act.

While a number of plasticizers used in PVC are regulated food additives (di-2-ethylhexyl adipate or DEHA is regulated without limitations in 21 CFR 178.3740), at least one, di-2-ethylhexyl phthalate, DEHP, is prior sanctioned. The prior sanction in this case is limited to use on foods of high water content.

Further adding to the complexity of the issue is the fact that cling films, which in the past were used by consumers only to wrap food, but which are now frequently being used to cook or reheat food, have traditionally been considered to be "housewares." Housewares are exempt from the requirement for pre-market safety evaluation to which commercial food packaging is subject. This does not mean that "housewares" are not regulated at all by FDA. While (like prior sanctioned materials) they are considered exempt from pre-market regulation under Sec. 409 of the Federal Food, Drug and Cosmetic Act, they are still subject to Sec. 402, the so-called General Safety Clause. FDA can still take action with respect to a "houseware", but the burden of proof is on FDA to prove that the food contacted by the "houseware" has been adulterated.

The Ministry of Agriculture, Fisheries and Food (MAFF) in the U.K. has expressed concern about the levels of plasticizers migrating from PVC film used in microwave applications. Research on migration of plasticizers from PVC film was conducted by MAFF's Food Science Laboratory in Norwich. Their results were published in 1987 as part of the Twenty-first Report of the Steering Group on Food Surveillance. What were described as "high levels" of plasticizer (DEHA) were found to migrate to foods with fat on the surface including spare ribs, chicken, trout, biscuits, cakes, bread and steamed pudding, when these foods were microwaved in direct contact with the PVC film. Lower migration levels were observed for situations where the film and the food were not in direct contact.

The report concluded that although the present estimated human dietary intakes are unlikely to represent any hazards to health, cling films should not be used in conventional ovens or microwave ovens where the film is in close contact with the food during cooking. A follow-up report, Food Surveillance Paper Number 30, was published in 1990.

The later report described substantial progress in lowering the exposure to DEHA. Manufacturers in the U.K. had switched to mixed-plasticizer cling films which contained substantially lower levels of DEHA together with polymeric plasticizers. Despite this, MAFF continued to recommend that cling films not be used in conventional ovens or for wrapping food or lining dishes when cooking in the microwave. MAFF further recommended that consumers not use cling films for wrapping food with a high fat content such as cheese.

There is a considerable body of data that indicates that both DEHA and DEHP are animal carcinogens. There has been a substantial amount of research conducted to demonstrate that these substances are secondary carcinogens (they induce peroxisome proliferation in the liver). The validity of the secondary mechanism argument and the issue of whether a secondary mechanism exempts a substance from the Delaney Clause are still being discussed.

On Dec. 14, 1990, FDA met with representatives from industry, SPI and CMA to request information on current uses of plasticizers in cling films in the U.S. This information will enable FDA to update its exposure estimates for the various plasticizers.

As part of its Total Human Exposure Research Program, the U.S. Environmental Protection Agency has funded a contract with Midwest Research Institute (MRI) to study migration from packaging materials commonly used in the home for heating foods in microwave ovens. The study will consist of the following two phases:

Phase I - To identify containers and packaging materials commonly used in the home during microwave heating of foods. Prioritize the potential for the identified containers and packaging materials to contaminate food based on material composition and anticipated usage.

Phase II - To conduct simulated household microwave heating studies using model food and beverage systems for those materials of highest priority to determine migration of chemicals into foods.

MRI has completed Phase I of their study. They developed a set of factors to be used for prioritizing the materials to be studied. Among the factors are the following: (1) frequency of use of the material, based on a survey which MRI conducted, (2) available toxicity information on the potential migrants (3) amount of migration data in the literature for the particular packaging material (4) the number of components and/or decomposition products which could potentially migrate from a given type of material.

Based on their prioritization scheme, MRI identified two types of packaging as highest priority for their Phase II migration studies: cling wraps and non-disposable plastic containers. Samples from these two categories will be used in their initial migration studies. FDA will be kept informed of their work as it progresses.

So far, I have been focusing on the fact that while food packaging regulations are written generically with respect to manufacturer, they are tied to a particular range of use conditions. This fact has important consequences as technology advances. Not only can we expect to see regulations for new food packaging materials, but also amendments to cover new uses of regulated materials.

For the remainder of my talk, I would like to focus on a second principle inherent in the regulations for indirect food additives. For the most part, these regulations assume that food packaging materials are made from a controlled source. I say for the most part, because there is a regulation in 21 CFR Part 176, which covers paper and paperboard, which does address, in a very general way, the use of other than virgin materials in the manufacture of paper food contact articles. The regulation (21 CFR 176.260) is entitled Pulp from reclaimed fiber, and it provides for the use of both industrial waste from the manufacture of paper and paperboard as well as salvage from used paper and paperboard provided that neither "bears or contains any poisonous or deleterious substance which is retained in the recovered pulp and that migrates to the food . . . ."

Part 177, which covers polymeric materials, has no comparable regulation. There is, however, a general regulation which is applicable to all food-contact materials, 21 CFR 174.5. Subparagraph (a)(2) requires that:

Any substance used as a component of articles that can contact food shall be of a purity suitable for its intended use.

The fact that the regulations for food contact plastics do not include any specific authorization for the use of recycled materials should not be interpreted to mean that FDA intended or intends to prohibit their use. As I have said a number of times in speeches to audiences like this one, FDA is not opposed to the use of recycled materials in food packaging. Rather, I suspect that the explanation for the absence of a regulation addressing the use of recycled plastics in food packaging applications is two-fold: (1) FDA did not anticipate this technology when the regulations in Part 177 were being written and (2) no one petitioned the Agency to establish such a regulation.

Recycling technology is now very much with us, however. While FDA does not intend to prohibit its use in food-contact applications, the Agency has the same responsibility vis a vis this technology as it does with any other: to ensure the safety of the U.S. food supply. In order to do that FDA needs to clearly establish specific criteria for the safe use of recycled plastics in food packaging applications. 21 CFR 174.5 does not contain any specific guidance as to what constitutes suitable purity. For that matter, the existing regulation on pulp from reclaimed fiber is very general and could be interpreted as requiring a zero tolerance for contaminants that could migrate to food. Clearly, some clarification is called for.

FDA intends to publish a regulation or regulations that will establish the requirements for the use of recycled plastics in contact with food. FDA is also developing protocols for industry to use in order to demonstrate that a particular recycled material is of suitable purity for food-contact use.

In demonstrating suitable purity, there are two types of potential contaminants that will need to be addressed. One type consists of residual amounts of the original contents of the package. The other type arises from incidental contamination resulting from materials other than the original contents being stored in the package prior to its being recycled. It is this latter type of contamination that is the focus of FDA's concern both because there is the potential for some very toxic materials (pesticides, gasoline, etc.) being stored in the package prior to recycling, and because the universe of this type of potential contaminant is very large.

Because the universe of potential contaminants is so large, it would not be reasonable or possible for industry to address them one by one. Rather FDA's protocol will call for model contamination studies with a representative group of contaminants of different polarities, volatilities and permeabilities to the polymer. The aim of the studies will be to demonstrate that the recycling process is capable of removing all but insignificant levels of the matrix of model contaminants.

Obviously, it will be necessary to define "insignificant levels" in the protocol so that industry will know how hard to look. As a general rule the level of contamination of recycled material which will be considered to be insignificant will be that which corresponds to a dietary exposure no greater than 0.5-1.0 ppb. Because the definition of insignificant will be given in terms of dietary exposure, the acceptable level of contamination in the recycled material will be polymer-specific. In estimating exposure, one considers both migration levels and the fraction of the diet packaged in that particular material (the so-called consumption factor). Consequently, for materials like polyolefins, which are used extensively in food packaging, the maximum acceptable contaminant residue level in the recycled polymer will be lower than that for materials like PET or PVC which have lower consumption factors.

There are other potential safety issues associated with recycling polymer resins or actual packages for food use that will need to be addressed. The regulatory status and the stability of the adjuvants in the polymeric resin will have to be considered. If packages are to be simply cleaned and reused, then the issue of durability in terms of package integrity will be important.

I hope that I have been able to provide you with some insights into how advances in technology will influence the kinds of regulations we are likely to see in the near future. These are innovative times for the food packaging industry. The rapid pace of technology makes it all the more important that we continue to work together to ensure that food safety is never compromised.

LEGISLATIVE INITIATIVES AFFECTING PLASTIC PACKAGING FOR FOODS

Glenn Gamber
Director, Program Management
Process/Package Technologies Division
National Food Processors Association (NFPA)
1401 New York Ave., N.W.
Washington, DC 20005

<u>CURRENT REGULATORY INITIATIVES ON PLASTIC PACKAGES</u>

INDUSTRY'S STAKE IN THE MARKET AS REGULATOR

PRESENTED BY

GLENN GAMBER

DIRECTOR, PROGRAM MANAGEMENT

NATIONAL FOOD PROCESSORS ASSOCIATION

at The Plastics Institute of America's 8th Annual Conference,

FOODPLAS '91, Orlando, Florida, March 5-7, 1991

GOOD AFTERNOON. I'M GOING TO COVER SOME GROUND WITH YOU WHICH HAS ALREADY BEEN TOUCHED ON BY SOME OF THE OTHER SPEAKERS ON THE PROGRAM, BUT I'M GOING TO PUT WHAT I THINK IS A SLIGHTLY DIFFERENT SPIN ON IT.

THE BIGGEST CHALLENGE FACING PACKAGING OF ALL TYPES TODAY IS THE SOLID WASTE ISSUE AND THE DIFFERENT WAYS IN WHICH THIS ISSUE MAY MANIFEST ITSELF IN FEDERAL, STATE, AND LOCAL LAWS AND REGULATIONS.

WHAT I'D LIKE TO SUGGEST TO YOU IS THAT HOW GOVERNMENT RESPONDS TO THE SOLID WASTE ISSUE TOUCHES ON SOME FUNDAMENTAL PRECEPTS OF YOUR BUSINESS. INDUSTRY, PARTICULARLY THE FOOD INDUSTRY, MUST BE VERY AWARE OF THE LARGER ISSUES AT STAKE. IT MUST SEE THE FOREST FOR THE TREES.

SOME LAWS AND REGULATIONS HAVE ALREADY BEEN PASSED. OTHERS HAVE OR WILL BE GIVEN SERIOUS CONSIDERATION. LET ME PROVIDE A BRIEF SURVEY OF THE KINDS OF ACTIVITIES GOVERNMENT IS ALREADY INVOLVED IN:

*CONEG SOURCE REDUCTION COUNCIL - THE SOLID WASTE PROBLEM IS PARTICULARLY CRITICAL IN THE NORTHEAST SO IT IS NO SURPRISE THAT THIS HAS BEEN A REGION WITH CONSIDERABLE ACTIVITY. THE SOURCE REDUCTION COUNCIL OF THE COALITION OF NORTHEASTERN GOVERNORS HAS DONE A NUMBER OF THINGS IN THE SOLID WASTE AREA:

-MODEL TOXICS LEGISLATION, PASSED IN A NUMBER OF NORTHEAST STATES AND SOME OUTSIDE OF THE NORTHEAST, LIMITS THE AMOUNT OF HEAVY METALS THAT CAN BE INTENTIONALLY USED IN PACKAGING.

-"PROTOCOL FOR THE IDENTIFICATION OF HEALTH AND ENVIRONMENTAL EFFECTS OF PACKAGING MATERIALS IN MUNICIPAL SOLID WASTE": THIS 70-PAGE DOCUMENT IS BEING REVIEWED BY A WIDE VARIETY OF INTERESTS AND DEALS, AS THE NAME IMPLIES, WITH AN APPROACH TO DETERMINING WHAT, IF ANY, OTHER SUBSTANCES SHOULD BE SUBJECT TO LIMITS IN PACKAGING.

-PREFERRED PACKAGING MANUAL - STILL IN DRAFT FORM, THIS LENGTHY DOCUMENT SETS OUT PREFERENCES IN PACKAGING AIMED AT ELIMINATING OR MINIMIZING PACKAGING, INCREASING REFILLABLE OR REUSABLE PACKAGING, AND INCREASING RECYCLABLE PACKAGING OR PACKAGING UTILIZING RECYCLED CONTENT.

-CHALLENGE TO INDUSTRY - REPORTEDLY BY THE END OF THIS MONTH, THE NINE NORTHEAST GOVERNORS WILL BE WRITING TO THE TOP 200 USERS AND PRODUCERS OF PACKAGING TO PUBLICLY CHALLENGE THESE COMPANIES TO PURSUE THE APPROACHES EMBODIED IN THE PREFERRED PACKAGING MANUAL BY SETTING GOALS AND REPORTING TO THE SOURCE REDUCTION COUNCIL ON PROGRESS TOWARD MEETING THOSE GOALS.

*STATES - THERE ARE VOLUMES OF BILLS IN THE STATES RELATED TO SOLID WASTE. THEY CENTER AROUND THE IDEAS OF PACKAGING TAXES, WASTE GENERATOR FEES, ADVANCE DISPOSAL FEES, RECYCLED CONTENT REQUIREMENTS, PACKAGING BANS AND/OR THE CONCEPT OF ENVIRONMENTALLY ACCEPTABLE PACKAGING.

-OSPIRG - THIS REFERENDUM INITIATIVE, DEFEATED LAST FALL, WOULD HAVE BANNED PACKAGES WHICH DID NOT MEET CERTAIN RECYCLING LEVELS, CONTAIN CERTAIN HIGH LEVELS OF RECYCLED CONTENT, OR WERE CAPABLE OF BEING REUSED FIVE TIMES. WHILE THIS WAS DEFEATED, IT IS

RESURFACING IN LEGISLATIVE PROPOSALS IN SOME STATES.

   -<u>MAINE</u> - HAS BANNED ASEPTIC PACKAGING.

   RCRA - THE REAUTHORIZATION OF THE RESOURCE CONSERVATION AND
RECOVERY ACT IS NOT NOW EXPECTED TO MOVE QUICKLY THROUGH CONGRESS.
INSTEAD, HEARINGS ARE EXPECTED TO BE HELD ON THE ISSUE THIS SPRING
WITH MARKUP OF A BILL POSSIBLY IN THE FALL.   HOWEVER, FULL
CONGRESSIONAL CONSIDERATION IS NOT EXPECTED UNTIL LATE NEXT YEAR.
IN FACT, INDICATIONS ARE THAT THE REAUTHORIZATION MAY NOT EVEN
HAPPEN IN THIS CONGRESS.   NO COMPREHENSIVE BILL HAS YET BEEN
INTRODUCED, BUT WHEN ONE IS THERE IS LIKELY TO BE AN EMPHASIS ON
POLLUTION PREVENTION AND A CONTINUING EMPHASIS ON THE ROLE OF THE
STATES IN DEALING WITH THE SOLID WASTE PROBLEM.

   ENVIRONMENTAL LABELING - AS YOU'VE ALREADY HEARD TODAY, THREE
STATES HAVE GOTTEN INTO THE ENVIRONMENTAL LABELING ARENA -
CALIFORNIA, NEW YORK, AND RHODE ISLAND.   CALIFORNIA'S LAW WOULD
REGULATE REPRESENTATIONS OF "OZONE FRIENDLY", "BIODEGRADABLE",
"PHOTODEGRADABLE", "RECYCLABLE" OR "RECYCLED".   A BILL IN THE
LEGISLATURE NOW WOULD ADD "COMPOSTABLE" TO THAT LIST.

   NEW YORK'S REGULATION APPLIES TO USE OF THE TERMS "RECYCLED",
"RECYCLABLE", OR "REUSABLE" AND CREATES STATE EMBLEMS FOR THESE
TERMS WHICH MANUFACTURERS MAY WISH TO USE.   NEW YORK HAS MADE CLEAR
THAT ITS REGULATION DOES NOT APPLY TO THE USE OF THESE WORDS IN
OTHERWISE TRUTHFUL STATEMENTS.

   RHODE ISLAND REGULATIONS COVER "RECYCLED", "RECYCLED CONTENT",
"RECYCLABLE", AND "REUSABLE" AND CREATE STATE EMBLEMS FOR EACH OF

THESE. UNLIKE NEW YORK'S ALLOWANCE FOR USE OF THESE WORDS IN OTHERWISE TRUTHFUL STATEMENTS, RHODE ISLAND PURPORTS TO REGULATE "ANY COMBINATION OR DERIVATION" OF THESE TERMS. STANDARDS INTENDED TO BE THE BASIS FOR REGULATIONS BY OTHER STATES HAVE BEEN ISSUED BY THE NORTHEAST RECYCLING COUNCIL AND THEY CONTAIN SOME OF THE ELEMENTS OF THE RHODE ISLAND REGULATION.

NFPA HAS FILED A PETITION WITH THE FEDERAL TRADE COMMISSION SEEKING NATIONALLY UNIFORM ENVIRONMENTAL MARKETING GUIDELINES WHICH I WILL DESCRIBE FURTHER LATER. SEN. FRANK LAUTENBERG OF NEW JERSEY HAD A BILL IN CONGRESS LAST YEAR, WHICH HE PLANS TO REINTRODUCE, WHICH WOULD REQUIRE EPA TO ESTABLISH SPECIFIC STANDARDS FOR ALL ENVIRONMENTAL CLAIMS.

IN THIS ARRAY OF ACTIVITY THERE ARE THINGS WHICH INDUSTRY MIGHT CONSIDER APPROPRIATE, BUT SOME OTHER THINGS WHICH INDUSTRY SHOULD CLEARLY CONSIDER INAPPROPRIATE ACTION BY GOVERNMENT.

GENERALLY, ENVIRONMENTAL ISSUES HAVE FOCUSED ON ONE OF TWO AREAS - ENVIRONMENTAL PROTECTION - I.E. IMPACTS ON PLANTS, WILDLIFE AND THE LIKE; AND HEALTH AND SAFETY - I.E. IMPACTS ON WATER, AIR, FOOD WHICH RELATE TO HUMAN WELL BEING. AS WITH MANY OTHER ENVIRONMENTAL MATTERS, THE SOLID WASTE ISSUE HAS BECOME A CONCERN AS OUR SOCIETY'S SENSITIVITIES INCREASE AND AS WE GATHER MORE INFORMATION ABOUT ENVIRONMENTAL AFFECTS EITHER ON THE ENVIRONMENT OR HUMAN HEALTH AND SAFETY. IT WAS NOT TOO LONG AGO THAT THE RESPONSE TO DECLINING LANDFILL CAPACITY WOULD HAVE BEEN TO SIMPLY BUILD MORE LANDFILLS. TODAY, THIS IS NOT A SIMPLE PROCESS. IN SOME CASES IT

IS IMPOSSIBLE. REGULATORY REQUIREMENTS FOR LANDFILLS ARE STRINGENT, THE PUBLIC RESISTS LANDFILLS NEAR THEIR NEIGHBORHOODS, AND INSURANCE CARRIERS ARE RELUCTANT TO PROVIDE NEEDED COVERAGE.

WITH LANDFILLING LIMITED AS A SOLID WASTE DISPOSAL OPTION, WE ARE EXPERIENCING A SHORT TERM SHORTAGE IN OUR CAPACITY TO EFFECTIVELY AND SAFELY DEAL WITH OUR SOLID WASTE. SOLID WASTE FROM PACKAGING HAS RAISED CONCERNS OF BOTH ENVIRONMENTAL PROTECTION AND HEALTH AND SAFETY. NEITHER OF THESE SHOULD BE LONG TERM CONCERNS. INDEED, FOOD PACKAGING CAN AND SHOULD TAKE FULL CREDIT FOR THE CONSIDERABLE ROLE IT PLAYS IN CONTRIBUTING TO THE HEALTH AND SAFETY OF AMERICAN CONSUMERS. AS WE DEVELOP NEW CAPACITY TO DEAL WITH SOLID WASTE, RELYING MORE HEAVILY ON APPROACHES OTHER THAN LANDFILL, PACKAGING, PARTICULARLY FOOD PACKAGING, WILL RETURN TO ITS RIGHTFUL PLACE IN CONSUMER'S EYES AS A MAJOR CONTRIBUTOR TO THE STANDARD OF LIVING WE ENJOY. BEFORE REACHING THAT POINT REDUCTIONS IN PACKAGING WILL OCCUR. RECYCLING WILL INCREASE. COMPOSTING AND WASTE TO ENERGY INCINERATION WILL INCREASE. AND THESE THINGS WILL ALL OCCUR IN A MANNER COMPATIBLE SOUND ENVIRONMENTAL PRACTICES.

IN THE MEAN TIME, WE MUST RESIST EFFORTS TO PERMANENTLY CHARACTERIZE PACKAGING AS A LONG TERM ENVIRONMENTAL PROBLEM AND WE MUST DO EVERYTHING WE CAN TO BE SURE THAT THE MARKETPLACE IS THE PRIMARY REGULATOR OF OUR PACKAGES AND THE PRIMARY STIMULANT OF POSITIVE ENVIRONMENTAL CHANGE. GOVERNMENT SHOULD ENCOURAGE MARKETPLACE DRIVEN ENVIRONMENTAL IMPROVEMENTS, NOT CREATE MARKETPLACE DISRUPTION BY WAY OF PACKAGING BANS OR INAPPROPRIATE,

ARBITRARILY STRINGENT ENVIRONMENTAL REQUIREMENTS. SUCH GOVERNMENT ACTION WILL ONLY DIVERT RESOURCES FROM INVESTMENTS IN ENVIRONMENTALLY BENEFICIAL CHANGES. CONSUMERS AND IN TURN INDUSTRY ARE RESPONDING AND WILL CONTINUE TO RESPOND TO THE NEED FOR CHANGE WITHOUT EXTENSIVE GOVERNMENT INTERVENTION.

ON FEBRUARY 14, THE NATIONAL FOOD PROCESSORS ASSOCIATION, JOINED BY 10 OTHER CO-PETITIONING ORGANIZATIONS, PETITIONED THE FEDERAL TRADE COMMISSION TO ADOPT VOLUNTARY INDUSTRY ENVIRONMENTAL MARKETING GUIDES FOR CLAIMS OF RECYCLED CONTENT, RECYCLABILITY, SOURCE REDUCED, COMPOSTABLE, AND REFILLABLE/REUSABLE. THIS WAS THE CULMINATION OF EIGHT MONTHS OF DILIGENT EFFORT BY OUR MEMBERS. THE PETITION ALSO REFLECTED INPUT BY THE CO-PETITIONING ORGANIZATIONS.

THIS STEP BY INDUSTRY - AND IT WAS A BROAD CROSS SECTION OF INDUSTRY WHICH JOINED IN THE PETITION - REPRESENTED SOMETHING OF A WATERSHED IN THE EVOLUTION OF THE SOLID WASTE ISSUE. WHEN THE PROCESS FIRST BEGAN, CONEG AND OTHERS WERE TAKING A VERY PRESCRIPTIVE APPROACH TO THE WHOLE AREA OF SOLID WASTE RELATED ENVIRONMENTAL LABELING. RECYCLING RATES AND RECYCLED CONTENT LEVELS WOULD HAVE HAD TO BE ACHIEVED BEFORE THESE TERMS COULD BE USED. MUCH OF INDUSTRY'S THINKING WAS FRAMED BY THE CONEG EFFORT.

THERE IS NOW A DIFFERENT ENERGY DRIVING THE ISSUE IN INDUSTRY. RESPONSIBLE COMPANIES WANT TO ACT RESPONSIBLY. THEY JUST WANT ONE SOURCE OF GUIDANCE. AND RESPONSIBLE COMPANIES REALIZE THAT THEY ARE GOING TO HAVE TO BE PART OF THE SOLUTION, WORKING WITH

COMMUNITIES, THROUGH THEIR TRADE ASSOCIATIONS, AND IN THEIR OWN PRACTICES TO REDUCE WASTE, STIMULATE RECYCLING, AND PRODUCE PRODUCTS AND PACKAGES COMPATIBLE WITH SOUND SOLID WASTE MANAGEMENT. AND THEY WANT TO COMPETE ON THIS BASIS.

THE SOLID WASTE PROBLEM IS ONE THAT IS PARTICULARLY SUSCEPTIBLE TO COMPETITIVE MARKETPLACE SOLUTIONS. LET ME EMPHASIZE AGAIN THAT THE PRINCIPLE NEED IN THE SOLID WASTE AREA IS FOR ENVIRONMENTALLY SOUND CAPACITY TO DEAL WITH OUR SOLID WASTE. WE NEED MORE RECYCLING INFRASTRUCTURE, MORE COMPOSTING CAPACITY, AND MORE WASTE TO ENERGY INCINERATION, AND ENVIRONMENTALLY SOUND LANDFILLS.

OUR PETITION ASKS FOR FTC GUIDANCE, NOT PRESCRIPTIVE REGULATION OF THE USE OF THE FIVE TERMS. WE ENVISION THE FREE FLOW OF INFORMATION TO CONSUMERS DO THAT THEY WILL ACT AS A MARKET FORCE FOR PRODUCTS REPRESENTING GENUINE ENVIRONMENTAL IMPROVEMENTS AND A STIMULUS TO INCREASED RECYCLING, COMPOSTING AND THE LIKE.

ONE OF THE OFTEN CITED CONCERNS IN THIS AREA, IS THAT TERMS USED IN ENVIRONMENTAL MARKETING ARE TOO TECHNICAL AND THUS SHOULD BE HANDED OVER, AS SEN. LAUTENBERG SUGGESTS, TO THE EPA WHERE THERE IS EXPERTISE TO DEAL WITH THE ISSUE. WE DISAGREE. THE STEPS WHICH WILL HAVE TREMENDOUS IMPACT ON THE PROBLEM ARE TRADITIONAL BUSINESS CONCERNS - MANUFACTURING PRACTICES -HOW TO DO THE SAME JOB WITH LESS MATERIAL OR WITH RECYCLED MATERIAL; MARKETS - HOW TO DEVELOP AND STIMULATE MARKETS FOR RECYCLED MATERIAL; AND MARKETING - HOW TO DEVELOP PACKAGING AND PRODUCTS WHICH REPRESENT GENUINE

247

ENVIRONMENTAL IMPROVEMENTS AND WHICH APPEAL TO CONSUMERS' INTENSE INTEREST IN THE ENVIRONMENT. THE EXPERTISE TO ACHIEVE THESE THINGS IS RIGHT WHERE IT ALWAYS HAS BEEN, IN THE FREE ENTERPRISE SYSTEM AND THAT SYSTEM WORKS BEST WITH THE MAXIMUM NUMBER OF COMPETITORS, NOT BY GOVERNMENT ARBITRARILY SKEWING COMPETITION IN WHAT IT CONSIDERS TO BE THE RIGHT DIRECTION.

SOME COMPANIES HAVE EXPERIENCED SHORT TERM GAINS AS THE RESULT OF PRECIPITOUS GOVERNMENT ACTIONS SUCH AS PACKAGING BANS. NO ONE SHOULD TAKE PARTICULAR COMFORT IN A COMPETITORS' MISFORTUNE. THE ENVIRONMENT IS A BIG ISSUE. FOR PACKAGE MANUFACTURERS IT HAPPENS TO BE FOCUSED NOW ON SOLID WASTE. IT MAY FOCUS ON OTHER THINGS AT A LATER TIME AT WHICH POINT TODAY'S LOSERS COULD BE TOMORROW'S WINNERS. WE FEEL STRONGLY THAT THERE SHOULD BE NO WINNERS OR LOSERS OTHER THAN THOSE DETERMINED BY THE MARKETPLACE. FULL BLOWN COMPETITION AMONG ALL MATERIALS AND ALL TYPES OF PACKAGING WILL PRODUCE THE GREATEST NET ENVIRONMENTAL IMPROVEMENT.

GOVERNMENT AND INDUSTRY SHARE AN IMPORTANT AND CRITICAL RESPONSIBILITY IN MAKING SURE THE MARKETPLACE WORKS, THAT IS IN EDUCATING CONSUMERS. GOVERNMENT AND INDUSTRY MUST TOGETHER EDUCATE THE PUBLIC SO THAT THE MARKETPLACE WILL BEHAVE IN AN ENVIRONMENTALLY RESPONSIBLE WAY. ONE DOES NOT NEED TO BE A PH.D. SCIENTIST TO UNDERSTAND BASIC FUNDAMENTAL CONCERNS OF THE SOLID WASTE ISSUE. BUSINESS PEOPLE DO NOT UNDERESTIMATE CONSUMERS AND THEIR ABILITY TO UNDERSTAND THE DIFFERENCE BETWEEN TRUTHFUL AND MISLEADING INFORMATION. NEITHER SHOULD THE GOVERNMENT. THE

MARKETPLACE WILL DETERMINE WHAT IS GOOD AND BAD PACKAGING. GOVERNMENT - BE IT FEDERAL, STATE, OR LOCAL - SHOULD NOT MAKE THIS DECISION.

LET ME CLOSE WITH SOMETHING OF AN EXAMPLE OF THIS PRINCIPLE. ONE THING CONSUMERS UNDERSTAND VERY CLEARLY IS FOOD SAFETY. ASSURING FOOD SAFETY INVOLVES GOOD MANUFACTURING PRACTICES, QUALITY CONTROL PROCEDURES, PROPER HANDLING DURING DISTRIBUTION AND SALE, AND PROPER HANDLING OF FOOD IN THE HOME. AT EACH STEP IN THIS PROCESS THERE IS EXPERTISE WHICH MUST BE APPLIED, YET A FUNDAMENTAL UNDERSTANDING AND APPRECIATION OF FOOD SAFETY IS WELL WITHIN CONSUMERS' GRASP. FOOD SAFETY IS A KEY MISSION OF NFPA WHICH IS WHY IN THE CONTEXT OF CONCERN OVER THE SOLID WASTE ISSUE WE ARE EXAMINING HOW WE MIGHT CONTRIBUTE TO SOLUTIONS IN THIS AREA WITHOUT COMPROMISING FOOD SAFETY. SPECIFICALLY THE NATIONAL FOOD PROCESSORS ASSOCIATION IS IN THE FORMATIVE STAGES OF DEVELOPING RESEARCH WHICH WOULD CONTRIBUTE TO CLEARING THE WAY FOR THE USE OF RECYCLED PLASTIC AS PRIMARY FOOD CONTACT PACKAGING IN APPLICATIONS WHERE SAFETY CAN BE DEMONSTRATED. WE ARE WORKING WITH THE SOCIETY OF THE PLASTICS INDUSTRY AND ARE CONSULTING WITH FDA TO FORMULATE APPROPRIATE RESEARCH. THIS IS AN AREA WHICH MUST BE APPROACHED VERY CAUTIOUSLY. AS YOU NO DOUBT KNOWN, THERE ARE FOOD CONTACT APPLICATIONS OF PLASTIC PACKAGING CONTAINING RECYCLED MATERIAL TO WHICH FDA HAS GIVEN ITS BLESSING, SPECIFICALLY FOAM EGG CARTONS AND PET SODA BOTTLES. BUT THE AGENCY HAS MADE IT CLEARLY PUBLICLY AND TO US THAT IT FEELS THE NEED, AS DO WE, TO BE CAUTIOUS IN THIS AREA. WE MUST NOT LET THE PUBLIC BELIEVE THAT BECAUSE RECYCLED

MATERIAL CAN BE USED IN ONE CASE THAT IT CAN BE USED IN ALL CASES. THE VARIABLES THAT AFFECT WHETHER OR NOT RECYCLED MATERIAL CAN BE SAFELY USED FOR FOOD CONTACT ARE NUMEROUS. THE ANSWER TO WHETHER OR NOT IT CAN BE SAFELY DONE WILL LIKELY VARY FROM CASE TO CASE. WE WILL BE LOOKING AT WAYS TO DETERMINE QUALITY CONTROL PROCEDURES OR TESTING METHODS WHICH CAN BE EMPLOYED TO ASSURE THE SAFETY OF CERTAIN BROAD CATEGORIES OF USES OF RECYCLED MATERIAL IN CONTACT WITH FOOD. THIS IS A CONSIDERABLE CHALLENGE WHICH WE HOPE TO BEGIN SOON.

THE FREE ENTERPRISE SYSTEM SHOULD BE ALLOWED TO FUNCTION BECAUSE IT CAN AND WILL RESPOND. OUR CHALLENGE IS TO CHARGE AHEAD AND TACKLE THE PROBLEM AND IN SO DOING ASSURE IT IS GIVEN THE OPPORTUNITY TO RESPOND RATHER THAN REACT TO ILL ADVISED GOVERNMENT ACTION, NO MATTER HOW WELL INTENTIONED.

THANK YOU.

SESSION 5

DESIGN FOR PLASTICS PACKAGING

DESIGN CONSIDERATIONS FOR PLASTICS PACKAGING
Frank E. Semersky
Senior Associate
Plastics Technologies, Inc.
333 14th Street
Toledo, OH 43624

Biography

Frank E. Semersky holds a Bachelor's degree in Biological Sciences from Columbia University and has fifteen years of varied experience in the design of medical devices and in the packaging of reagents and perishables for both the food and medical industries. Frank is the author or co-author of nine U.S. patents. His educational background and extensive experience in all phases of food packaging in both monolayer and multilayer plastic containers and in shelf life and accelerated testing methods uniquely qualify him to understand a customer's package design requirements based on oxygen and flavor barrier analysis and on the requirements dictated by the packaging process itself, including bottle handling, filling, sealing or capping, and processing systems. Frank's plastic processing experience includes injection molding, extrusion blow molding, statistical process control, and Taguchi methodology. He also has extensive international and domestic project management experience and serves as Director of Operations for PT Health Care Products, a PTI joint venture. Frank was a 1988 winner of the SPHE Packaging Award Competition in the Medical Device Category for his design and commercial development of a novel tamper-resistant container for biological specimens.

DESIGN FOR PLASTIC PACKAGING

by

Thomas E. Brady, Ph.D.
President, Plastic Technologies, Inc.

and

Frank E. Semersky
Senior Associate, Plastic Technologies, Inc.

Safe ... hot fillable ... retortable ... long shelf life ... microwaveable ... dual-ovenable ... convenient ... tamper evident ... fresh ... home made taste ... shelf appeal ... priced right ... environmentally friendly ... The "hot buttons" in food packaging keep getting more numerous. How can package designers and package users be certain that they are concentrating on the right issues and how can they can be certain that an investment made today in either a new package design or in implementing a new package design will pay for itself in the future? Certainly, the array of packaging and consumer priorities and the endless chase to find a niche in the marketplace makes this question almost impossible to answer. What then should a designer, manufacturer or user do? How should choices be made? What is the right package for the product?

While there may never be simple and straightforward answers to these questions, nevertheless, it does seem that the "hot buttons" can be prioritized in a way which may help to distinguish package design basics from those attributes which provide an opportunity for differentiation. For example, while all of the attributes listed above may have some measurable influence on whether a package is right for the product, only two of the suggested attributes on that list will be used to judge a package choice on a Pass/Fail basis. Those two attributes include Safety and Price. Consumers will not purchase an item if it is even perceived to be unsafe. Certainly no food processor in this room would attempt to offer for sale anything which was unsafe, but if the package even suggests less than perfect regard for safety, the consumer is turned off. Simple things like collapsed panels, dents, wrinkled foil seals and tamper bands broken during application, for instance, all give the appearance of potentially unsafe contents while, in fact, the contents are perfectly acceptable by any measure of food safety.

Marketing studies have shown that consumers have a price threshold for a given product, no matter what the package option. Therefore, if you must bet on future viability of a package, especially for a food product, you had better not neglect to take the cost of the package into account even if it is better than sliced bread. Thus, the two attributes, Safety and Cost, can make or break a package design.

A second decision level can be defined around the attribute of quality. Once consumers can have safety and economy, they will tend to make the next level of judgement on the basis of quality. Why is this true, and why would quality win over convenience? The answer to this question is that convenience is a more relative attribute. Consumers have some influence over the perception of convenience of any given package and, in general, it has been well established that consumers will pay for convenience but that they will opt for less convenience if quality becomes a factor. For example, while two different easy-open, dual ovenable packages may be judged to be quite different in terms of "peelability", nevertheless if both packages can be opened but the package which opens more easily is judged to be poorer in quality, the consumer invariably will opt for the higher quality product next time around. The consumer has inherently less influence over quality and he/she knows it, so if all other attributes are not totally unacceptable, quality becomes the leading factor in consumer decisions, particularly for repurchase.

So where do convenience, environmental friendliness and shelf appeal stand as attributes? In our judgement, convenience still comes in ahead of either shelf appeal or environmental factors. This is not to say that either of these other two factors does not have an important influence and that environmental factors might not be forced to be a higher priority decision based on governmental edicts about recycling. However, convenience is a more explicitly definable attribute and once the consumer finds that a package is safe, economical, delivers a product high in quality, and is convenient as well, he/she has already identified it for repurchase. Shelf appeal may buy a product some initial influence but repurchase will be decided based on other factors.

But how about environmental issues? Isn't it true that every consumer poll seems to indicate that people are willing to buy recyclable and environmentally acceptable packages over other options? The answer to this question is indeed a resounding "yes." However, the practical fact is that you can't eat or drink an environmental package. Therefore, this type of attribute becomes a secondary influence in the actual purchase of a product. The key point is that the other factors ahead on the decision making list must be at least close to equal. If they are equal, then indeed environmental attributes become the deciding factors.

Where does this leave package designers, manufacturers and users today as they face the world of package choices? Unfortunately, it means that none of the attributes can be completely ignored when designing packages since the array of available options gives most consumers what we might define as "reasonable" choices in the market place. However, it does mean that in order to get a payback on your investment in packaging, the "fundamentals" must be inherent in the package design; that is, safety and economy (better defined as competitive price) must be achievable. Once safety and economy are assured, the priority must be product quality and only after these attributes are assured should the priorities shift to convenience and then finally to environmental friendliness and/or to such attributes as shelf appeal.

I realize this line of thinking may appear to be out of sync with what we read about environmental awareness today, but remember that ten years ago these same priorities fit package design and at that time, environmental issues were not even considered to be a major factor. Our advice is to stick to the basics and to use the other attributes to achieve more marketable packages and to create some "differentiation" in the market place.

The ideal strategy, of course, is to design packages that have all the basic attributes and which are also paragons with regard to all the extra attributes. Perhaps this is possible but, a more realistic scenario is that we need to examine the realities of designing the ideal package or the "right package for the product at the right price" by considering where the state-of-the-art is with regard to package design attributes.

Indeed, that is exactly what you are going to hear as we move through this session on "Design in Plastic Packaging." Today you will hear several experts deal with the attributes of Safety as it relates to shelf life and product protection; Quality as it relates to freshness and product formulations for today's microwaveable packages; Convenience as it relates to Easy Open packages, and Environmental Friendliness as it relates to making monolithic packages work. We will conclude with a discussion which attempts to define what we mean by Environmentally Acceptable performance.

Now, let's sit back and enjoy what I think you will agree are several very interesting and thought-provoking discussions of these topics.

"BARRIER PACKAGING TECHNOLOGIES -   WHAT ARE THE ALTERNATIVES?"

Ronald H. Foster
Technical Development Manager
EVAL Company of America
1001 Warrenville Rd., Suite 201
Lisle, IL 60532

Biography

Ron Foster is the Technical Development Manager for Eval Company of America. He has over 20 years experience in all phases of polymer production, applications and markets. He is a Senior member of the Society of Plastic Engineers, a Professional member of The Institute of Packaging Professionals and is on the Board of Trustees of the Plastics Institute of America.

Abstract

The paper will present an overview of the barrier polymers and technologies available today. Details of technologies will be presented where possible. Future projections will also be presented.

BARRIER PACKAGING TECHNOLOGIES
WHAT ARE THE ALTERNATIVES?

by

RONALD H. FOSTER
TECHNICAL DEVELOPMENT MANAGER
EVAL COMPANY OF AMERICA

INTRODUCTION:

Without a doubt, the most significant development in food packaging
over the past 15 - 20 years has been the use of barrier polymers and
barrier packaging technology.

With the growth that multilayer plastic barrier packaging has exhibited
in the last six years, and the projected future growth for this type of
packaging, the rush is on to develop new barrier resins, films and
technologies. Many polymer companies have developed new "barrier"
materials while other companies, already producing barrier resins, are
extending their offerings with new and improved barrier resins and
barrier technologies.

To put all of these new developments into perspective is a difficult
task. Many of the new resin technologies are not yet commercially
available, so the only data available is that supplied by the
manufacturer. This paper will provide an overview and, where possible,
details of these developments.

DEFINITION:

We have heard a lot about barrier packaging, but what is a barrier?
According to The American Heritage Dictionary, a barrier is
"something that hinders or restricts". If we apply this definition to
barrier packaging, then we have packaging that either prevents
undesirable substances from reaching the packaged product, or prevents
desirable substances from leaving the packaged product. These
substances may be oxygen or other gases, flavors, aromas, moisture or
solvents.

Given this definition, two of the alternatives for barrier packaging
must then be metal and glass. This paper however, concerns itself with
the "non-traditional" forms of packaging such as barrier polymers
and/or barrier technologies. To define barrier polymers/technologies,
I would like to paraphrase Mr. Morris Salame of Polysultants Company.

"A polymer or technology which, when used alone or in combination with non-barrier layers, protects the product over its intended long-term ambient shelf-life by virtue of its barrier to gases and organic flavors and odors, and is characterized by an O2 P-Factor of < 10 (cc-mil)/(100 in^2 - d-Bar)".

HISTORY:

The use of plastic barrier packaging predates the 1960's. Figure 1 is a chronology of plastic barrier packaging in the last thirty years.

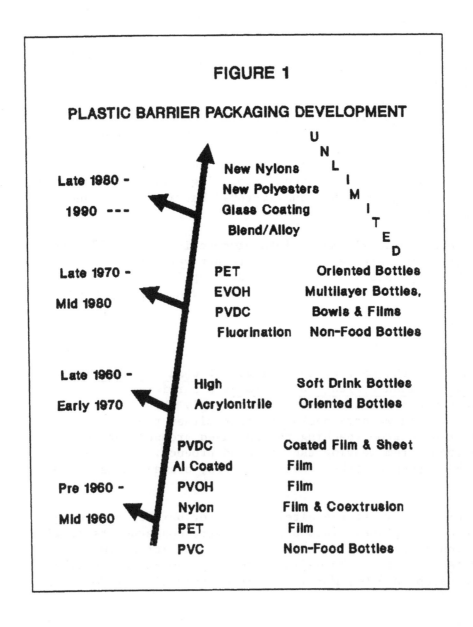

FIGURE 1

PLASTIC BARRIER PACKAGING DEVELOPMENT

It wasn't until the last ten years however, that the predominance of new polymers and/or technologies have been introduced. In general, barrier packaging during the 1970's and 1980's has concentrated on multilayer, multi-material structures. For a variety of reasons including environmental, economics, etc., we are now seeing more and more work in the areas of monolayer containers, coatings, alloys, etc.

BARRIER POLYMERS:

Barrier polymers or resins used for food packaging must combine both oxygen and moisture barrier properties. They can be divided into either high barrier or moderate barrier categories according to the following definition.

High Barrier Polymer

Oxygen Transmission Rate: 1.0 cc.mil/100 in^2.day.atm @ 20°C, 65%RH
Water Vapor Transmission Rate: 10 g.mil/100 in^2.day.atm @ 40°C, 90%RH

For other packaging applications, barrier properties to solvents, chemicals, heat, etc., may be the prerequisite.

HIGH BARRIER POLYMERS:

The two most widely used high barrier polymers for food packaging are Ethylene Vinyl Alcohol Copolymers (EVOH) and Polyvinylidene Chloride (PVDC). Both of these polymers are available throughout the world and offer excellent resistance to the permeation of gases, flavors, aromas, solvents and most chemicals. PVDC is also an excellent barrier to moisture while EVOH offers very good processability and the ability to use large amounts of regrind in the food package.

Worldwide there are five suppliers of EVOH copolymers. They are:

o   EVAL® Resins (Kuraray Co. Ltd., & EVAL Company of America (EVALCA)

o   Soarnol® Resins (Nippon Goshei)

o   Clarene® Resins (Solvay)

o   Selar OH® (Du Pont)

EVOH resins are produced in a wide variety of grades ranging from a 27 mol% ethylene content to a 48 mol% ethylene content. Generally speaking, the lower the ethylene content, the higher the barrier to gases, flavors, solvents, etc. EVOH resins are normally used in coextrusions with polyolefins, nylon or polyethylene terephthalate (PET) as the structural layers.

PVDC resins are available as extrudable resins, latexes, soluble resins and micronized powders by the following suppliers;

o   Saran® Resins (Dow Chemical)
o   Emulsions (Union Chemicals)
o   Ixan® Powder and Coating Resins (Solvay)

They are used in coextrusions, as coatings for various substrates and in monolayer or coextruded films. Although PVDC has excellent barrier properties to gases and moisture, they are difficult to process and are corrosive.

In addition to EVOH and PVDC the other high barrier polymers are:

o   Barex®

     This is an acrylonitrile copolymer produced by BP Chemicals, Inc.
     Barex® resins exhibit excellent machinability, high gloss, high
     stiffness, medium gas barrier properties and excellent resistance
     to a wide range of chemicals.

     Barex® resins are only approved for direct non-beverage food
     contact up to fill temperatures of 150°F (65°C). Due to this
     limitation, the only food related applications commercially using
     Barex® resins are processed meat related.

o   Barrier Nylon

     MDX6 Nylon, produced by Mitsubushi Gas Chemical Company, Ltd, is
     a crystalline resin produced through the polycondensation of
     meta-xylene diamine with adipic acid. In addition to Mitsubushi,
     MDX6 is also available from Solvay and Toyobo. This material
     has good gas barrier properties, higher tensile values, higher
     glass transition temperature and less water absorption than
     Nylon 6 or 66. Commercially, MDX6 is being used with PET for
     coinjection blow molded containers and in PET/MDX6 blends for
     films and bottles.

o   Amorphous Nylon Resins

     There are two general types of Amorphous Nylon resins available.
     One, Novamid X21 produced by Mitsubishi Chemical Industries Ltd.,
     is a copolyamide comprised of hexamethylene diamine and
     isophtalic acid. The other type, Selar PA® produced by DuPont
     and Gelon® A-100 produced by General Electric Company are both
     polycondensation polymers consisting of isophthalic acid,
     terephthalic acid and hexamethylenediamine.

     Amorphous nylon resins are characterized by high clarity, high
     strength properties, a lower water absorption rate than Nylon 6
     and fair to good chemical resistance. The gas barrier properties
     of amorphous nylon resins fall right at the borderline between
     high and moderate barrier resins.

Commercially, amorphous nylon resins are being promoted for monolayer bottles and films. To date, the only known commercial applications are a monolayer bottle for macademia nuts and coextruded polycarbonate/amorphous nylon refillable soft drink and water bottles in Europe.

o   Liquid Crystal Polymers

The first commercial LCP was DuPont's Kevlar® which was based on alyotropic aramid composition.

Both Vectra®, produced by Hoechst Celanese, and Xydar®, produced by Amoco Chemicals, are thermotropic liquid crystal polymers.

LCP's are based on aromatic polyester chemistry and exhibit a highly ordered molecular structure in both the molten and solid states. This results in aggregations of rigid, rodlike molecules which align themselves in parallel fashion during molding. This creates a self-reinforced structure with high mechanical properties.

LCP's have exceptionally high heat distoration temperatures, good chemical resistance, excellent moisture barrier properties and very good gas barrier properties. Reportedly, the price of LCP's is in the $10 per pound range which will probably preclude their use in food packaging. Target markets mentioned have been automotive, electronics, industrial and communication applications.

Table I compares various strengths and weaknesses of these high barrier polymers.

MODERATE BARRIER POLYMERS

According to the previous definition, a moderate barrier polymer can include any polymer that has an oxygen transmission rate greater than 1.0 cc.mil or an WVTR of greater than 10 gm.mil or a combination of both. This of course would include all of the polyolefins and most other polymers we could think of. In this paper however, we will only address those polymers that have some unique properties that might be applied to packaging.

o   Polyethylene Terephthalate (PET)

There are many new developments in the area of barrier PET. Mr. Muir of Goodyear Tire and Rubber will discuss these in a subsequent paper so they will not be addressed here.

o   Polyacrylic Imide

Kamax® is an amphorous acrylic-imide copolymer produced by Rohm-Haas.
This material offers the traditional properties associated with
acrylics, such as clarity, surface hardness and weatherability. It
also has high heat resistance and a high modulus. Oxygen and water
barrier properties are in the fair range. Kamax® is presently being
used in the automotive industry and for some optical lens
applications.

o   Polyetherimide (PEI)

In 1982, General Electric introduced a family of high temperature
amorphous thermoplastic polyetherimide polymer under the trade name
Ultem®.

Some of the important characteristics of these polymers include high
heat resistance, high flexural strength, low mold shrinkage,
transparency, excellent electrical properties, and a broad resistance
to oils, chemicals, and other fluids. It has excellent hydrolytic
stability, good microwave distribution characteristics, and is
resistant to essentially all food products.

One promising application for PEI is coextruded film and sheet for
food packaging. The primary use would be in disposable dual-ovenable
frozen food trays. Dual-ovenable capability refers to trays that can
be heated in both microwave and in conventional convection ovens. This
requires a 425-450°F working temperature and microwave resistance.
PEI has the added quality of evenly distributing microwaves.

o   Polyvinyl Alcohol

Vinex® is a thermoplastic copolymer of polyvinyl alcohol and
poly (alkyleneoxy) acrylate. It is produced by Air Products and
Chemicals. Although Vinex® is still cold water soluble, it is
extrudable either by injection molding, cast film, tubular blown film,
or injection blow molding. Vinex® resins are also truly biodegradable.

Although polyvinyl alcohol has an exceptionally high barrier to gases,
Vinex® resins, due to the addition of a plasticizer, have only
moderate gas barrier properties. It has very good barrier properties
to greases, oils and non-aqueous concentrated chemicals.

It is not known if Vinex® has found any commercial applications.
Potential applications cited are as a lamination with paperboard to
form a biodegradable, heat sealable package for greases, oils or
non-aqueqous substances, injection blow molded bottles to package
concentrated pesticides, herbicides or fertilizers and as flexible
packages for dry products.

o <u>Polymethylpentene (PMP)</u>

   PMP resins are produced by Phillips 66 Company (Crystalor®) and
   Mitsui Petrochemical Industries (TPX®). They are crystalline
   polyolefins that are characterized by outstanding clarity, excellent
   heat and chemical resistance and good resistance to greases and oils.
   PMP resins also have very low moisture absorption characteristics.
   For the near future, Phillips foresees the greatest packaging
   potential for these resins in monolayer containers requiring moderate
   gas barrier properties.

o <u>Polyvinyl Chloride (PVC)</u>

   In 1986, the USFDA withdrew their 1975 proposal banning PVC bottles
   and sheet in food-contact applications. The new proposal sets
   guidelines for maximum acceptable levels of VCM monomer.

   PVC is known for its transparency and chemical resistance. It also
   has moderate barrier to gases. PVC is being used for household
   chemicals, edible oils, toiletries, cosmetics and medical/health
   care applications. In 1985, PVC ranked third among all rigid
   packaging materials.

Table II compares the moderate barrier polymers.

<u>BARRIER FILMS</u>

There are a number of unique barrier packaging films available today. Most
of these films are based on the barrier polymers discussed previously. The
following is an overview of the films available.

o <u>Biaxially Oriented EVOH</u>

   Biaxially oriented EVOH films, such as Kuraray's EF-XL film, combine
   the inherent barrier properties of EVOH with biaxial orientation
   technology to produce a unique packaging film. These films exhibit
   superior mechanical and optical properties. The use of the biaxial
   orientation and heat setting technology reduces the moisture
   sensitivity of these films while greatly enhancing the dimensional
   stability. These films are used in laminations, can be metallized
   or used as a monolayer film for some applications.

o <u>Biaxially Oriented Nylon</u>

   Allied-Signal is the only U.S. producer of biaxially oriented nylon
   films. Their capacity is reported to be approximately nine million
   pounds per year.

   Biaxially oriented nylon offers better machinability and printing
   as well as higher strength and transparency than non-oriented films.
   These films are positioned as a competitor to PET and OPP films and
   are being used for packaging cheeses, meats, frozen foods, coffee
   and for liquid pouches.

o   Polyacrylonitrile Film (PAN)

PAN films are produced from high nitrile resins. Mobil Chemicals
has a patented technology for the production of these films. However,
they have chosen not to produce the films, but rather license the
technology. It is not known if anyone has chosen to take a license or
is even considering it. Biaxially oriented PAN film is characterized
by excellent clarity and high mechanical properties. Chemical
resistance and oxygen barrier properties are very good.

o   Saranex® Films

Saranex® films, produced by Dow Chemical, are coextruded of
polyolefins, usually LDPE, and Saran® (PVDC) resins. These films have
been used for many years, either by themselves or in laminations,
for chub packages, pouches, bag-in-box and form, fill seal
applications.

The barrier properties achieved depend upon the Saranex® grade however,
all would fall into the moderate barrier range. The cost of the
films is also rather high, ranging from 10 cents to 14 cents per MSI.

o   Oriented Polyester (OPET)

Biaxially oriented polyester is a high performance film used in a wide
range of packaging applications. The properties of OPET films include
heat stability, moisture resistance, clarity, machineability and
chemical resistance. Gas barrier properties are enhanced by metallizing
or coating with a barrier resin such as PVDC.

o   Silica Coating

Silica coated film is probably the newest barrier technology to be
developed. It has received more instant attention than any other
development in packaging history. There is some misconception as to
what these films really are. Silica is the common name for Silicon
Dioxide which is the primary component of glass. The molecule used to
coat films however, is not silica rather a variation called Silicon
Oxide (SiOx).

Although this technology is still in its infancy, two methods of
applying the coating have emerged. One is by electron beam and the
other by plasma deposition. Both of the methods are relatively
expensive. According to the literature, SiOx coated films are glass
clear and can increase the oxygen and moisture barrier properties of
the base film by 4 to 8 times. SiOx coated polyester is reported to
have the gas and moisture barrier properties equivalent to that of
metallized PET without the flex cracking problems.

The only known commercial applications are in Japan with films produced by Toppan Printing. Several other companies including Bonar Flexible Packaging, Toyo G.T., Eastapac and Airco Coating Technology have development programs for both film and bottle coatings.

Although other types of films such as biaxially oriented polypropylene and others may be considered barrier packaging films, time and space precludes covering them all.

Table III compares the properties of various barrier films.

Up to this point, we have discussed the various barrier polymers and films available for packaging applications. There are also a number of interesting barrier technologies either in use or being developed for packaging.

BARRIER TECHNOLOGIES

o   Oxygen Absorbers/Scavengers

    There are two general types of oxygen absorbing technologies:

            o   Chemical
            o   Enzyme

    The chemical technology is the more common type and is made up of three systems:

    -    The most well known is the use of metallic reducing agents such as various ferrous compounds, powdered iron oxide or metallic platinum. These systems scavenge oxygen within the food package and turn it into a stable oxide. The newest of the metallic types is the Oxbar® system from CMB Packaging in the United Kingdom.

    -    Non-metallic formulations have also been developed and are intended to alleviate problems with metal detectors or the potential for metallic tastes/odors being imparted to food products. These systems employ compounds, such as ascorbic acids (Vitamin C) and their associated salts.

    -    The newest system consists of organo-metallic molecules that have a natural affinity for oxygen. Oxygen molecules irreversibility lock onto molecules within the chemical system, and thereby, free oxygen is extracted from the surrounding environment. Flexibility is achieved by altering the specific molecular design of the chemical system to control the strength and rate of this extraction. Aquanautics Corporation, Emeryville, California, is developing this system using a technology originally intended for the Department of Defense.

In addition to chemical oxygen scavenging systems, there are a number of enzymatic oxygen-getting systems. For instance, alcohol oxidase has been detected on several strains of yeast grown on methanol. Some of these enzymes have been isolated and characterized. They have demonstrated oxygen absorbing activity in the presence of alcohol to form an aldehyde in hydrogen peroxide. In another example, oxygen is absorbed by adding water to a mixture of glucose and glucose oxidase. The deoxygenating agent requires addition of water from the outside atmosphere for activation, and therefore cannot be used effectively with low moisture foods, unless the water activity is substantial. The whole idea of oxygen absorbing systems is to reduce oxygen in the head space of a package. Generally, their effectiveness is used up rapidly and therefore, they have limited use.

o   Fluorination

This is a process where the inside of a blow molded plastic container is treated with a dilute blend of fluorine in nitrogen during the blowing cycle. The fluorine reacts rapidly with the hot polymer, replacing hydrogen atoms on the polymer chains. As a result of this reaction, the wall of the container is altered such that non-polar hydrocarbon solvents can no longer permeate rapidly through the container.

Fluorination does provide good barrier properties for most hydrocarbons however, it does not work for chlorinated solvents, oxygen, carbon dioxide or moisture.

To date, fluorination is being used for solvent containers and some in automotive. It is a relatively expensive process and can be somewhat dangerous.

Air Products has used a modified fluorination process to provide a barrier to flavor scalping in a soft drink syrup bottle for Coca Cola. It is not know if this has been commercialized, however.

o   Selar RB®

This is a family of modified nylon concentrates that are mixed in various proportions with a polyolefin like HDPE. They are produced by DuPont. The key to using these materials is actually in the fabrication stage. DuPont has developed processing technology for conventional blow molding equipment that results in a layered structure containing discontinuous over-lapping barrier platelets. This creates a "tortuous path" which provides barrier properties. To use Selar RB,® this technology must be licensed from DuPont.

Selar RB® has good barrier properties to a variety of hydrocarbons, agricultural chemicals, industrial chemicals and paint solvents. It does not qualify for uses involving contact with food products under applicable FDA regulations..

o   Blends/Alloys

This is an area where the imagination can run wild. Combinations of
materials such as,

- Polyimide/Liquid Crystal Polymers
- MDX6 Nylon/PET
- Styrenics/Nylon

and many others are being looked at for improved packaging properties.
Most of this work is still in the development stage and where it
will lead is anyone's guess.

CONCLUSION

This paper has attempted to provide an overview of the barrier packaging
technologies available today. For every technology discussed, there are
probably two or three others being considered.

Does one technology stand out head and shoulders above another? I believe
the answer to this is that it depends upon the packaging requirements. The
use of coextrusion technology, to combine materials with various properties
into a package meeting the desired shelf life requirements, is probably the
candidate today. The desire to reduce costs, address environmental issues,
improve properties, etc., is the driving force behind the development of new
technologies and the improvement of existing ones.

I believe it is safe to say that,

"BARRIER PACKAGING TECHNOLOGY IS A MOVING TARGET".

TABLE I

HIGH BARRIER POLYMERS

Polymer	Gas Barrier	Chemical Resistance	Processing	Cost/ Performance
EVOH	When properly protected offers excellent barrier properties at all conditions	Excellent to all acids, bases and solvents	Thermally stable no special equipment required	Excellent
PVDC	Barrier properties not moisture sensitive, MA copolymer comparable to EVOH at high relative humidities	Excellent to acids and alcohols, Poor to most bases, hydrocarbons and esters.	Thermally sensitive, equipment requires corrosion protection	Excellent
Acrylonitrile	Gas barrier properties are an order of magnitude, lower than EVOH or PVDC	Very good to all acids, bases and solvents	High viscosity, thermally sensitive material resulting in lower outputs	Very good, can be used in monolayer applications
Barrier Nylon	Barrier properties are affected by moisture. Overall they rank between Acrylonitrile & EVOH or PVDC	Good resistance to most chemicals and solvents	Processes similar to most nylons, must be dried	High cost moderate performance
Amorphous Nylons	Poorest barrier of the high barrier polymers	Only fair resistance	Slight advantage over other nylons	Fair to good can be used in monolayer applications
LCP	High	Very good	Not difficult	High cost necessitates speciality applications

TABLE II

MODERATE BARRIER POLYMERS

Polymer	Chemical Resistance	Temperature Resistance	Gas Barrier	Other
Polyacrylic Imide	Very good	Excellent	Fair	High clarity, high modulus, weather-ability
Polyetherimide	Very good	Excellent	Poor	Excellent electrical properties
PVOH	Excellent	Poor	Excellent when dry	Water soluble biodegradable
Polymethylpentene	Excellent	Excellent	Fair	Low moisture absorption, high clarity
PVC	Excellent	Fair	Fair	High clarity

TABLE III
BARRIER FILMS

Property		B.O. EVOH	B.O. Nylon	B.O. PAN	Saranex14	OPET	SiOx
Thickness, mils		0.6	0.6	1.0	2.0	0.5	0.5
Tensile Strength, Break, psi	MD	29,000	28,600	27,000	3,500	29,900	22,800
	TD	28,000	31,300	27,000	2,700	31,000	27,000
Elongation, Break, %	MD	100	90		400	100	140
	TD	100	90		550	95	60
Youngs Modulus, psi	MD	510,000	242,000	160,000	32,000	555,000	498,000
	TD	510,000	213,500	140,000	31,000	598,000	570,000
Water Vapor* Transmission Rate		2.6	16.8	0.2	0.2	3.0	0.01–0.08
Elemendorf Tear, g	MD	260	500	250	700	200	100
	TD	330	450	350	460	350	100
Haze, %		0.5	2.0	0.5	26	0.5	0.5
Gloss, 45°,%		95	85	90	55	95	95
Oxygen Transmission** Rate							
20°C, 65% RH		0.02	3.2	–	–	–	–
85% RH		0.06	9.0	0.05	0.5	8.0	0.02–0.14
100% RH		0.39	31.6	–	–	–	–

*   g/100 in^2.day @ 40°C, 90% RH

**   cc/100 in^2.day.atm

POLYESTERS FOR MONOLITHIC BARRIER FOOD PACKAGING

Matt C. Muir
Senior Research Chemist
Polyester Research and Development Division
The Goodyear Tire & Rubber Company
130 Johns Ave.
Akron, OH 44305

Biography

The author earned his B.S. in Chemical Engineering at the University of Massachusetts in Amherst in 1981. From then until 1988, he studied Polymer Science and Engineering at the University of Massachusetts, under the direction of Roger Porter, receiving his Ph.D. in 1988. He is currently working as a Senior Research Chemist in New Polymers R & D in the Polyester Division at Goodyear.

Abstract

Processing and structural variables for attaining good barrier properties using single-layer polyester articles are presented. PET is a high-barrier polymer by itself; low permeability can be enhanced by stretching (film or bottle applications) or by crystallization (thermoforming, CPET). A higher-barrier polymer is poly (ethylene naphthalate); under similar conditions of stretch ratio, its oxygen permeability is 1/5 that of PET. In addition, its tensile strength and modulus are significantly higher than that of PET. For both polymers, permeability increases slightly with molecular weight.

POLYESTERS FOR MONOLITHIC BARRIER FOOD PACKAGING

In preparing for the future, plastic food packagers must consider several factors in their efforts to grow their opportunities and to stay current with state-of-the-art concepts. Among the issues packagers will have to consider are trends in consumer lifestyle, manufacturing procedures, and environmental and regulatory concerns. Growth in plastic food packaging will be enhanced by the consumer's need for quick, portable, prepared foods. Manufacturing driving forces will be in the direction of shorter shelf-life, due to direct-to-market trends and the push for just-in-time delivery. Polymer packaging can face regulatory difficulties, due to the public perception that it is not recyclable. However, polyester food packaging is a viable option in meeting this challenge, due to an existing strong recycling infrastructure, and its environmental acceptance as opposed to other polymer systems. An ideal example of the recycling capability of PET is Goodyear's REPETE. Currently, REPETE is available for non-food contact applications, and it is expected quickly to close the recycle loop back into the same food contact container.

The agenda for this talk is as follows. First, I will survey briefly the PET package food applications. Next, I will move to a discussion of the bottle-blowing process and some technical concerns relevant to that process. Then, a brief discussion of PET morphology will follow, leading to data on the parameters that affect crystallization and the effect of crystallinity and other morphological features on the permeability of products made from PET. After this, I will describe PEN, the first generation of Goodyear's HP family of polyesters, and the processing concerns in utilizing PEN. Finally, I shall give a brief summary of the work presented here.

PET

The main polyester for food package applications is poly(ethylene terephthalate), PET. In addition to the familiar beverage bottle, using, for example, Goodyear's CLEARTUF, some food applications in which PET is making a strong contribution are in the shelf-stable market, using, for example, Goodyear's TRAYTUF, in hot-fillable or retortable containers, in specialty barrier containers (when a co-component is utilized), as an alternative to coextruded packages (CPET), and with the addition of an oxygen scavenger to reduce the level of contamination in a filled container.

In these competitive times, it is imperative to exact maximum value out of a material in any process, and the food packaging process is a prime example. In gaining the best return for a packaging material, the designer must understand how the structure of the material interacts with the process of manufacture to produce the desired properties. Important process-structure factors include process temperature, degree of stretching, the intrinsic viscosity (IV) of the polymer, which is related to its molecular weight, and heat-setting procedure, if any. Included in the structural aspects of great interest are the degree and type of crystallinity, which is controlled by process and molecular weight effects, and which affects strength, thermal properties, and permeability; oxygen, water vapor and carbon dioxide permeability, impact strength, and tensile strength, elongation, and modulus of the resultant packages.

Figure 1 outlines some processing concerns in the stretch-blow-molding bottle process. In the neck-sidewall transition zone, there is little stretch in the axial direction and one to 5x in the hoop direction. Throughout the sidewall, the stretch is 2x in the axial direction and 5x in the hoop. In the neck, uneven material distribution, or "necking", must be avoided for a bottle to have good appearance. Oriented (lamellar) crystallinity in the bottle sidewall is desirable for high impact strength. In addition, the pressures of carbonation are contained well with oriented sidewall crystallinity. These characteristics, necking and strain-hardening, are related to the natural draw ratio of the polymer.

In working toward property improvement, the key is the morphology of the polymer. PET is versatile, in that one can tailor the morphology to meet his packaging demands. For a high-barrier package with a high heat-deflection-temperature, thermoforming of a CPET material will produce a high-crystallinity, high-temperature, high-barrier package. For barrier applications where clarity is important, blow-molding orientation of a preform made with a high-clarity formulation of PET will give a clear bottle with high tensile strength.

Figure 2 is a simplified schematic representation of a semicrystalline polymer. In both pictures, crystalline regions are depicted as sections where the polymer chain segments are parallel. The amorphous regions are depicted with the chain segments in a less ordered conformation. On the left, thermal crystallization is represented. In this case, the polymer chain segments are aligned with the direction of the crystal, but the orientation of one crystalline region is independent of the orientation of another. The polymer chain segments in the amorphous regions follow a random-walk conformation and are not characterized by an average orientation. Note that individual polymer chains can pass through crystalline and amorphous domains, and from one crystal to another.

The diagram on the right depicts strain-induced crystallization (strain-hardening). In this case, the crystals are smaller than those resulting from thermal crystallization. The crystalline domains in strain induced crystallization also tend to follow an average orientation in the direction of the strain. The chain segments in the amorphous regions also are subject to molecular-level strain, resulting in the phenomenon of "amorphous orientation". Thermally-crystallized samples, for instance, CPET trays, are white opaque parts, due to light-scattering caused by the difference in amorphous and crystalline refractive indices and the size of the crystallite clusters (spherulites). In contrast, strain-hardened articles, for instance, bottles and some film products, often have high clarity, due to two factors: 1. the crystalline domains are smaller than the wavelength of light, and 2. refractive index difference between crystalline and oriented amorphous regions.

In thermal crystallization, the determining variables to consider are temperature and molecular weight, represented by the polymer's intrinsic viscosity (IV). Figure 3 depicts a typical crystallization curve for PET homopolymer, using the crystallization half-time plotted against temperature for three IVs. At temperatures approaching Tm, the melting point, crystallization is slow, due to a small thermodynamic driving force. As the temperature is decreased, the thermodynamic driving force is increased, and crystallization proceeds more quickly. As the temperature approaches Tg, the glass transition temperature, the crystallization rate is kinetically controlled, since chain motion is slowed. For most of the curve where a difference can be discerned, the lower-IV polymers crystallize more quickly than the higher-IV polymers.

In seeking to increase the use temperature of PET, heat-setting is an option, rendering PET a viable polymer for hot-fill applications. The heat-setting process consists of constrained annealing of oriented bottles. The effect that can be achieved is shown in Figure 4, which compares the hot-fill properties of heat-set bottles with a control. All were made from the same polymer. Heat-setting, combined with alterations in bottle shape and sidewall thickness, can increase use temperatures by 20C or more.

In contrast to thermal crystallization of unoriented amorphous samples, heat-setting showed a monotonic increase in crystallinity with annealing temperature (Figure 5). The increase in chain mobility with temperature is most likely the dominant mechanism here. The data are not so clear here, but as with the unoriented case, lower IVs crystallize more readily, again most likely because of chain mobility.

The dependence of permeability on crystallinity is shown in Figure 6. There is considerable scatter, but it is clear that increased crystallinity decreases permeability. Most models describe this with an additive function, in which the permeability of the crystalline domains is zero and the morphological contribution to permeability depends on the path that gas molecules must traverse around crystalline regions and through amorphous domains in the film.

Partly because of strain-induced crystallization, gas permeability decreases with orientation (Figure 7). The mechanism is not limited to crystallinity, however; some effect of oriented amorphous segments, increasing the gas molecules' traveling distance through the film, is likely.

In summary, PET is a versatile material. Judicious choice of polymer grades and processing conditions result in a variety of properties, from highly-crystalline ovenable trays, to high-clarity, strain-hardened bottles and films. Increased crystallinity and orientation decrease permeability; heat-setting and high crystallinity increase the maximum operating temperature.

PEN

Poly (ethylene naphthalate) (PEN) was synthesized as early as the 1960's. Attempts at commercialization have in the past been unsuccessful, but a renewed effort toward high-purity monomer, dimethyl 2,6-naphthalene dicarboxylate, has generated great interest in PEN as a commercial polymer. Chemically, PEN is of course similar to PET. There are differences, due to the naphthalene rings, which impart to PEN a stiffer backbone. Understanding the differences between PEN and PET in structure and processing is required in order to utilize it to its full potential. The morphological options (crystallinity and orientation) for working with PET also apply to PEN.

Figure 8 outlines some applications for PEN, which at Goodyear is the first generation of our HP polyesters. Its high Tg (120 C, vs. 70 C for PET) makes it a candidate for hot-fillable bottles and high-temperature dielectrics. Its high strength and modulus can be utilized in magnetic tape and fibers, especially tire cord. Its low permeability is valuable for specialty bottle and other packaging applications.

The bulk of the experimental work for studying the structure/processing/property relationship of PEN comes from statistically designed experiments. Film samples were equibiaxially stretched in a TM Long film stretcher at DevTech Labs (Amherst, NH). The independent process variables studied were stretching temperature (150, 160, and 170 C), stretch ratio (2x, 3x, and 4x), and intrinsic viscosity (IV) (.63, .72, and .93, corresponding to molecular weights [Mn] of 36 000, 44 000, and 65 000) Dependent variables measured were crystallinity, oxygen permeability, and strength, elongation, and modulus of the resultant films.

Crystallinity determination has a high random error, due to temperature fluctuations and uncertainty in analyzing the DSC traces, However, the effect of conditions on properties can be seen. All results reported are statistically significant within a 95% confidence limit. Although molecular weight and temperature have an effect, the variable most affecting crystallinity is stretch ratio. For instance, Figure 9 shows the effect of stretch ratio on crystallinity, as measured by heat of fusion, for three IVs at 150 C. It is seen that the bulk of the crystallization occurs in the initial stretch, with subsequent less sharp increase in crystallinity.

Figure 10 gives the influence of IV on crystallinity. It is seen that crystallinity decreases with IV and increases with temperature. Since decreasing IV and increasing temperature both lower the relaxation time of the polymer, it is hypothesized that the kinetics of thermal- and strain-induced crystallization are promoted by lower polymer relaxation times, rather than by a high concentration of entanglements.

Oxygen permeability transverse to the film plane was measured on a MoCon OxTran 10/50 instrument at 30C. While the effect of IV on PEN's permeability is similar to what is seen with PET, orientation results in an even greater improvement with respect to PET. Unoriented PEN shows lower gas permeability (higher barrier) than PET (3.6, vs. 12 cc*mil/100in2/day/atm), and orientation serves to improve further its barrier properties. Figure 11 compares the effect of stretch ratio on O2 permeability for PEN stretched at 150 and 170 C and for PET (2). Oriented PEN can show as little as 1/5 the O2 permeability of PET (interpolated to 1.2, vs. 6.0). Despite more strain-induced crystallinity, and possibly because of lower amorphous orientation, the PEN sheet stretched at the higher temperature exhibits higher gas permeability.

In Figure 12, we see the effect of IV on gas permeability for PEN samples stretched at two temperatures and compared with some prior PET data(2). It is seen that the permeability of PEN is as much as five times lower than PET. Oxygen permeability in PEN decreases slightly with decreasing IV, probably because of increased crystallinity. With process temperature, permeability increases slightly, perhaps because of less amorphous orientation.

Although the statistical correlations did not result in a predictive trend in tensile strength and elongation to break with respect to processing parameters, some rough quantitative values could be seen. Figure 13 gives a table of these results, comparing PEN to PET film. It is clear that the tensile strength of unoriented and oriented PEN film is higher than that of PET, and that the elongation to break of PEN is lower than that of PET. Since PET crystallizes more facilely than PEN, this behavior is likely due to amorphous orientation and stiffness of the PEN chain brought on by the bulky naphthalene rings.

In conclusion, PEN is a high-barrier polymer whose properties improve with orientation. Its high Tg and tensile properties make it useful in a wide variety of applications. With increasing stretch ratio, crystallinity increases and oxygen permeability decreases. With increasing process temperature, crystallinity increases but oxygen permeability increases. With decreasing molecular weight, crystallinity increases and oxygen permeability decreases. Oriented PEN's gas permeability is only 1/5 that of oriented PET. Its uniqueness must be considered in the design of any processes.

Summary

PET and PEN are versatile packaging materials, due to the ability to control end use properties by polymer and processing variables.

References:

1. Cheng, S.Z.D., and Wunderlich, B., Macromolecules 1988, 21, 789-797.

2. Swaroop, N., and Gordon, G., Polymer Engineering and Science 20(1), 78-81 (1980).

3. Perkins, W., Polymer Bulletin 19, 397-401 (1988).

Bottle Process

- Materials must process well from zero-stretch to high-stretch regions

- Necking

- Strain-hardening

- Natural Draw Ratio

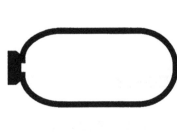

Morphology

- Crystallinity
  - Thermal
  - Strain-Induced

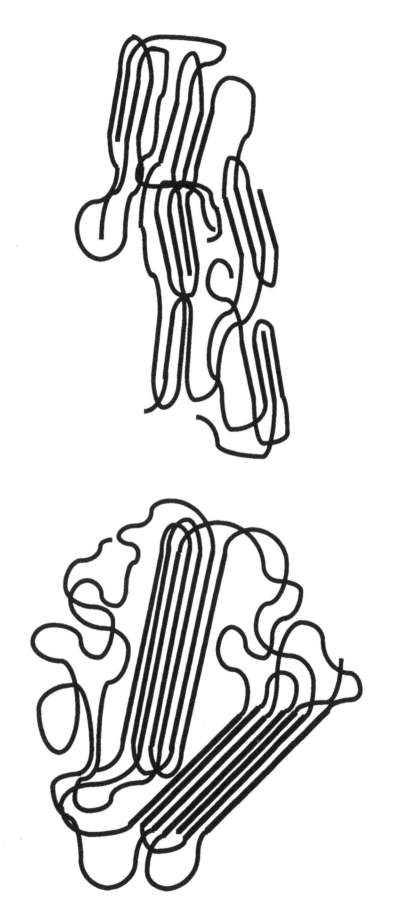

281

Tailoring Morphology

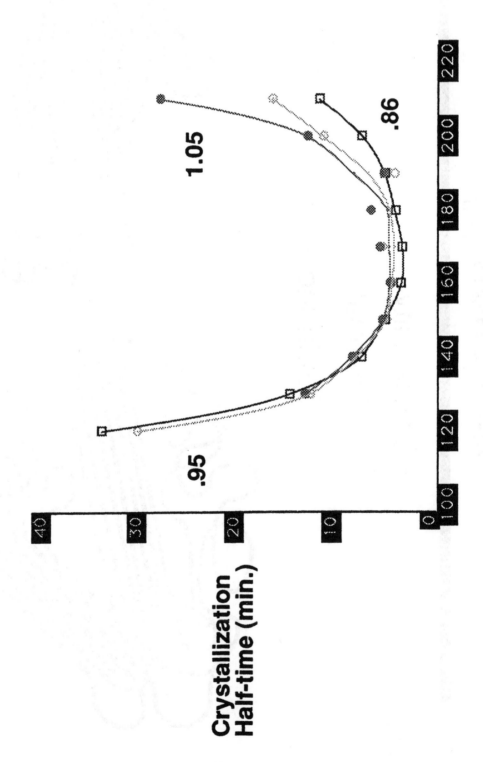

Crystallization Half-time (min.)

Temperature (°C)

1.05

.95

.86

Heat-Setting

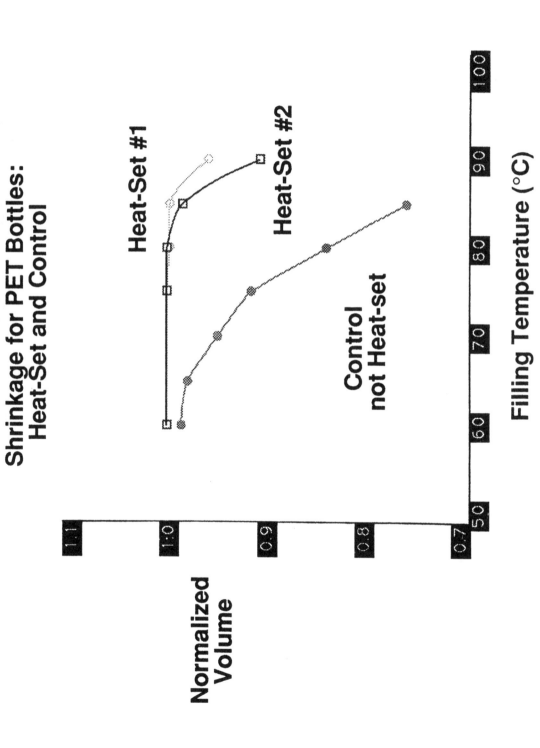

Shrinkage for PET Bottles:
Heat-Set and Control

Heat-Set #1

Heat-Set #2

Control
not Heat-set

Normalized
Volume

Filling Temperature (°C)

283

Effect of Annealing Conditions

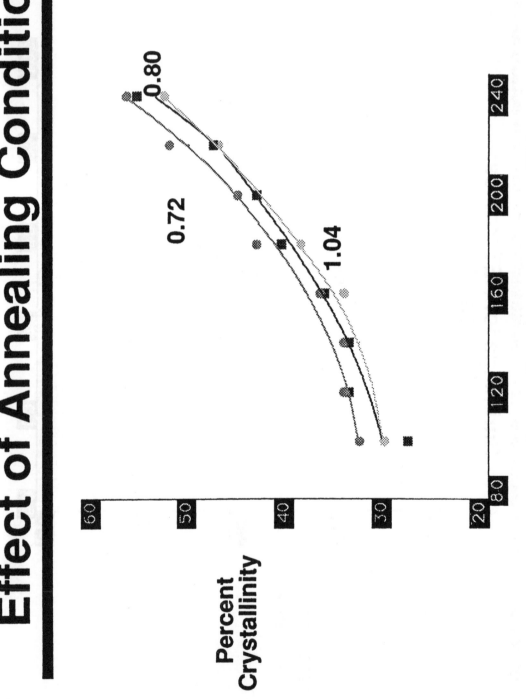

Percent Crystallinity

Annealing Temperature (°C)

0.80

0.72

1.04

Permeability vs. Crystallinity

Three IV's of PET

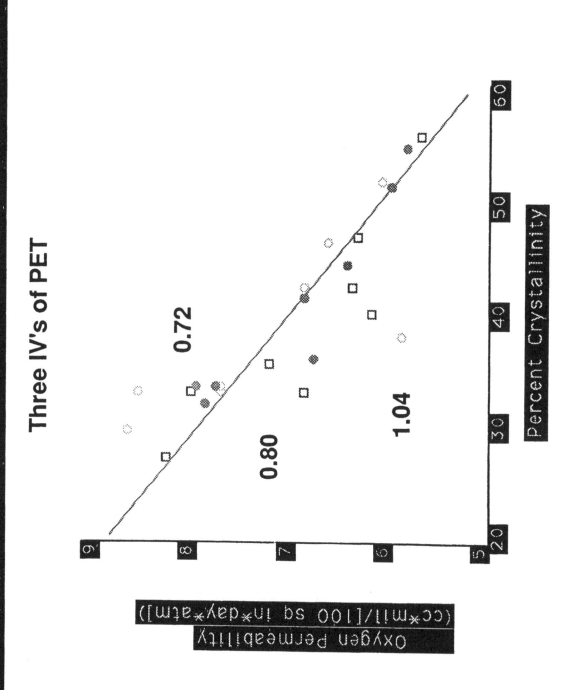

Permeability vs. Orientation

Not Heat-Set

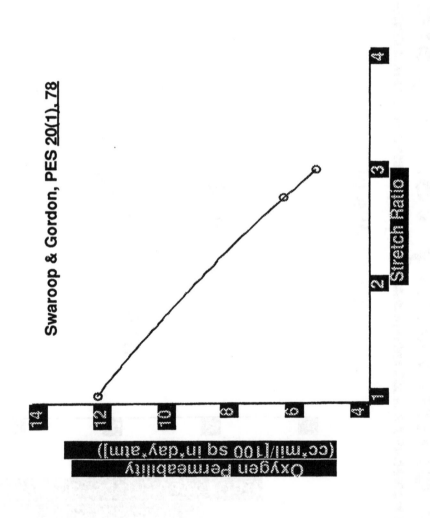

Swaroop & Gordon, PES 20(1), 78

Applications for PEN

- Bottles
  - Hot-fill (High Tg)
  - Specialty (Low Permeability)
- Film, Fiber
  - High Strength and Modulus
  - High Service Temperature

287

PEN
Crystallinity vs. Stretch Ratio

Stretched at 170°C, not Heat-Set

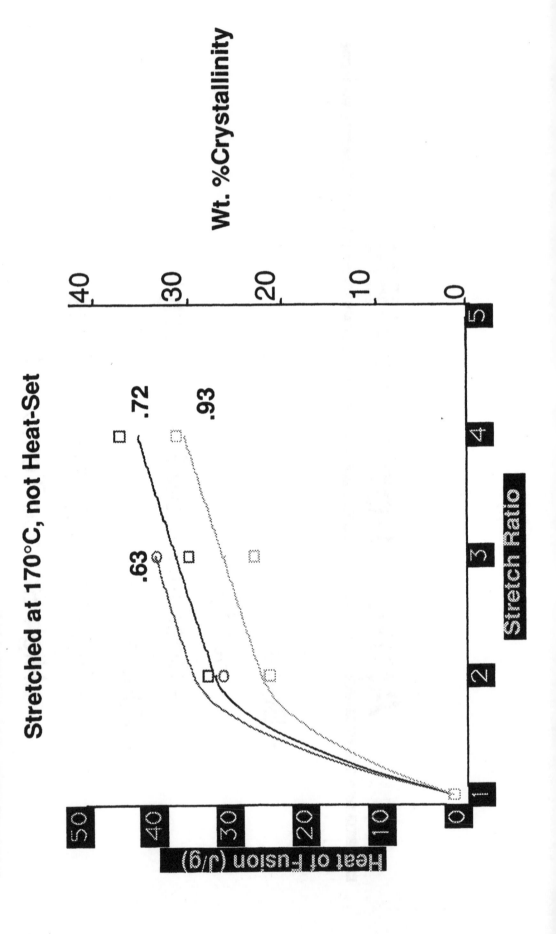

PEN
Crystallinity vs. IV

Stretch Ratio: 3x3, not Heat-Set

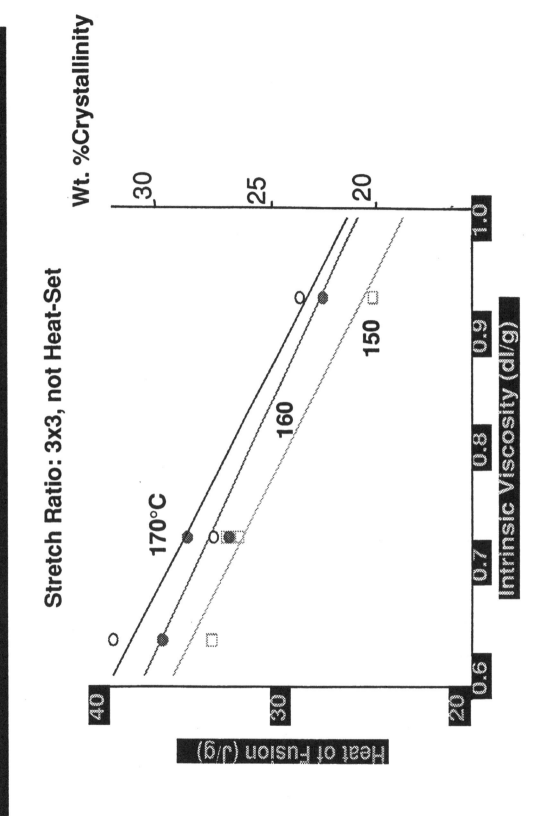

O₂ Permeability vs. Stretch Ratio

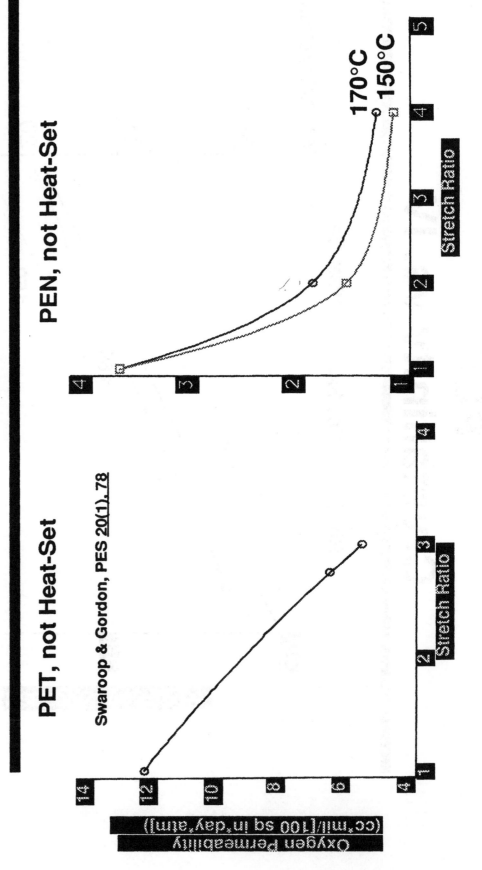

O$_2$ Permeability vs. IV

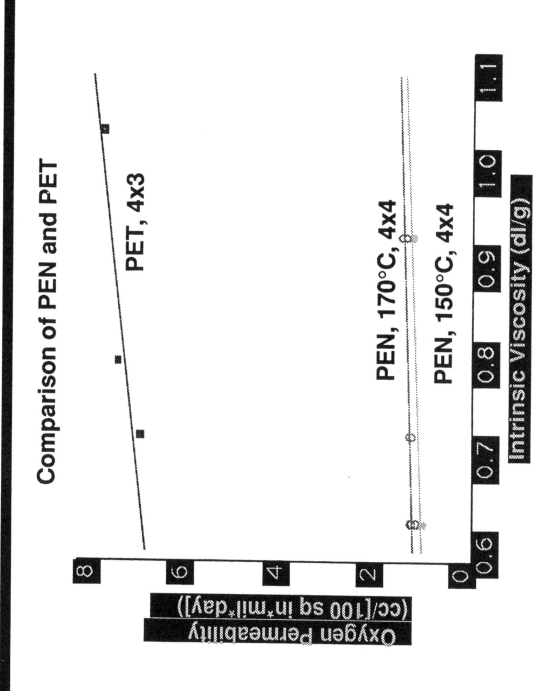

Comparison of PEN and PET

PET, 4x3

PEN, 170°C, 4x4

PEN, 150°C, 4x4

Intrinsic Viscosity (dl/g)

Oxygen Permeability
(cc/[100 sq in*mil*day])

Tensile Properties: PEN and PET

	PEN	PET
Tensile Strength, Amorphous	11 kpsi	7 kpsi
Oriented	25-60 kpsi	12-30 kpsi
Elongation to Break, Amorphous	200-280%	250-500%
Oriented	5-60%	10-85 %

CHILLED FOODS: PACKAGING FOR FRESHNESS

Cynthia Wilbrandt
Project Manager
A. Epstein International
600 W. Fulton St.
Chicago, IL 60661-1199

BIOGRAPHY

Ms. Wilbrandt is a Project Engineer with A. Epstein and Sons International, Inc., an engineering consulting organization which specializes in the design and construction of innovative food processing facilities. As a member of Epstein's Chilled Foods expert team, Ms. Wilbrandt is part of the team that is responsible for designing, engineering, and building facilities which minimize the risk of microbiological contamination and ensure the highest quality chilled food production.

Prior to Joining Epstein, Ms. Wilbrandt was Manager of Research and Development at Liquid Carbonic in Chicago, Illinois where she was responsible for establishing a modified atmosphere packaging technical support program which has been utilized by many food companies to develop safe and successful MAP packaging systems.

Ms. Wilbrandt has a B.S. in Chemical Engineering from Northwestern University, and has completed Silliker Laboratories short course in Food Microbiology. She has also spoken at numerous conferences on the topics of producing and packaging refrigerated, microbiologically sensitive foods.

ABSTRACT OF TALK

Fresh, preservative-free chilled foods require special attention in preparation and packaging to minimize microbiological contamination and maximize safe product shelf life.

Chilled food products must be prepared in a manner that will result in a product that is free of food pathogens such as Listeria Monocytogenes and Salmonella. Then, it is the function of the package to ensure product integrity. The packaging system for a chilled food must eliminate any opportunity of product contamination during the packaging step, and ensure that the product is protected from post-packaging contamination.

This talk will focus on the critical issues and operations involved in chilled food processing and packaging. This includes the process steps, how various technologies may affect packaging system requirements, the material handling precautions needed to minimize potential contamination, and an overview of the packaging methods and materials utilized by chilled food processors today.

CHILLED FOODS: ENGINEERING FOR PACKAGED FRESHNESS

OUTLINE OF TALK FOR FOOD PLAS '91

I.   INTRODUCTION

II.   REVIEW OF MICROBIOLOGICAL ACTIVITY IN CHILLED FOODS PRODUCT DEVELOPMENT

III.   CONTROLLING MICROBIOLOGICAL ACTIVITY

IV.   RAW INGREDIENT AND PACKAGING MATERIAL HANDLING AND STORAGE

V.   PROCESS TECHNOLOGIES

    A.   Assemble/Package

    B.   Cook/Chill/Assemble/Package

        1.   Temperature control

        2.   Clean rooms

        3.   Robotics

    C.   Cook/Chill/Assemble/Package/Pasteurize

        1.   Microwave pasteurization

        2.   Modified Clean Room

    D.   Cook/Chill/Assemble/Package/Sterilize

        1.   Microwave sterilization

        2.   Retort

    E.   Sous Vide

    F.   Hot Fill

    G.   Frozen and "Slacked Out"

CHILLED FOODS: ENGINEERING FOR PACKAGED FRESHNESS

FOODPLAS '91

MARCH 7, 1991

Cynthia S. Wilbrandt

A. Epstein and Sons International, Inc.

PRODUCT DEVELOPMENT

- FORMULATION

- PACKAGING

- PROCESSING

INTRINSIC PROPERTIES

- Acidity or pH

- Moisture content

- Nutrient content

- Occurrence of antimicrobial compounds

- Oxidation - Reduction potential

- Biological structure

EXTRINSIC FACTORS

- Temperature

- Relative Humidity

- Gaseous composition of environment

CHILLED FOOD PROCESSING

FOOD PROCESSING OPTIONS

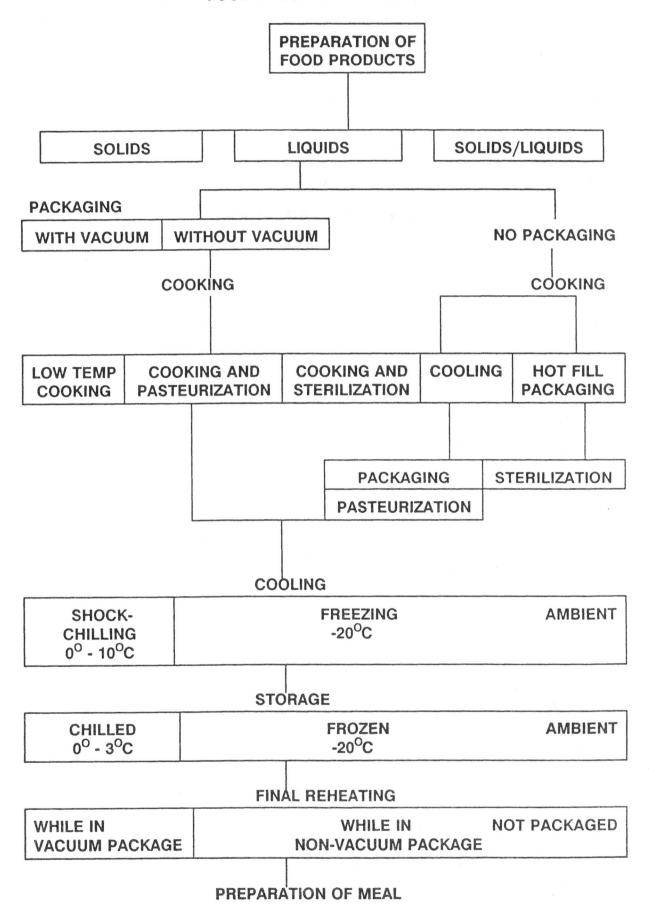

RAW INGREDIENT HANDLING

- Remove from primary package and place in sanitary container.

- Perform microbiological physical and chemical analysis.

- Store at appropriate temperature in dedicated storage.

- Minimize microbiological activity in storage.

- Utilize a first-in, first-out inventory system.

- Prevent cross contamination with cooked product.

- Only sanitary containers should enter processing operations, no wooden pallets or paper bags.

Assemble/Package

Assemble/Package

Cook/Chill/Assemble/Package

PROCESS TECHNOLOGIES

Assemble/Package

Cook/Chill/Assemble/Package

Cook/Chill/Assemble/Package/Pasteurize

PROCESS TECHNOLOGIES

Assemble/Package

Cook/Chill/Assemble/Package

Cook/Chill/Assemble/Package/Pasteurize

Cook/Chill/Assemble/Package/Sterilize

PROCESS TECHNOLOGIES

Assemble/Package

Cook/Chill/Assemble/Package

Cook/Chill/Assemble/Package/Pasteurize

Cook/Chill/Assemble/Package/Sterilize

Sous Vide

VACUUM COOKING PROCESS

CUSTOMER DRIVEN MENU

PROCUREMENT OF ONLY THE FRESHEST, HIGHEST QUALITY INGREDIENTS

PRE-PREPARATION OF PRODUCT

VACUUMIZED SEALING

COOKING UNDER VACUUM AND RAPID CHILLING

REFRIGERATED STORAGE AND JUST IN TIME DELIVERY

STORAGE AND RE-HEATING BY CUSTOMER

QUALITY ASSURANCE

↑

↓

PROCESS TECHNOLOGIES

Assemble/Package

Cook/Chill/Assemble/Package

Cook/Chill/Assemble/Package/Pasteurize

Cook/Chill/Assemble/Package/Sterilize

Sous Vide

Hot Fill

PROCESS TECHNOLOGIES

Assemble/Package

Cook/Chill/Assemble/Package

Cook/Chill/Assemble/Package/Pasteurize

Cook/Chill/Assemble/Package/Sterilize

Sous Vide

Hot Fill

Frozen and Slacked Out

PROCESSING EQUIPMENT

- ■ ADHERE TO STANDARDS (3-A, USDA, USPHS)

- ■ CORROSION FREE

- ■ EASY TO CLEAN

- ■ CONSIDER C.I.P.

- ■ "WHAT IF?" (ABUSE, LEAKS, WEAR)

Building Design:

- Sanitary construction and finish.

- HVAC   - Air balance
  - Filtration
  - Temperature control

- Appropriate floor drain specifications

- Plant layout should eliminate potential cross contamination from both a material handling and people flow design.

- Layout should isolate chilled assembly from other operations.

PACKAGING

Equipment:

- Must provide a reliable and consistent vacuum and/or gas flush.

- Must provide a reliable seal.

- Must be easily cleanable.

PACKAGING

Material:

■   Must be specified through appropriate testing.

■   Must be removed from outer covering and placed in a sanitary bag or container prior to entering packaging room.

■   May be surface treated prior to use.

HAZARD ANALYSIS CRITICAL CONTROL POINT (HACCP)

HACCP PRINCIPLES

1. Assess hazards and risks associated with the product from growth of raw ingredients to consumption of the food.

2. Determine the critical control point required to control the identified hazard.

3. Establish the critical limits that must be met at each identified critical control point.

4. Establish procedures to monitor the critical control points.

5. Establish corrective action to be taken when there is a deviation identified by monitoring a critical control point.

6. Establish effective recordkeeping systems that document the HACCP plan.

7. Establish procedures for verification that the HACCP system is working correctly. (Physical, chemical, sensory, or microbiological).

Critical Control Point Specifications:

■ The control point and its parameter.

■ The monitoring procedure

■ The monitoring frequency.

■ The decision criteria for acceptable and unacceptable control.

■ The action to take if control is unacceptable.

Packaging Recommendations

Of the Meat and Poultry Working Group of the National Advisory Committee and Microbiological Criteria for Foods, Mission Statement (1990).

1. Approved packaging materials are stored and handled in a manner consistent with maintaining microbial integrity.

PACKAGING RECOMMENDATIONS

1. Approved packaging materials are stored and handled in a manner consistent with maintaining microbial integrity.

2. Equipment sanitation programs, employee practices, and product handling procedures must minimize product contamination during the packaging process.

PACKAGING RECOMMENDATIONS

1. Approved packaging materials are stored and handled in a manner consistent with maintaining microbial integrity.

2. Equipment sanitation programs, employee practices, and product handling procedures must minimize product contamination during the packaging process.

3. Key factors such as product temperature, vacuum level, and modified atmosphere delivery systems, must be monitored and controlled.

PACKAGING RECOMMENDATIONS

1. Approved packaging materials are stored and handled in a manner consistent with maintaining microbial integrity.

2. Equipment sanitation programs, employee practices, and product handling procedures must minimize product contamination during the packaging process.

3. Key factors such as product temperature, vacuum level, and modified atmosphere delivery systems, must be monitored and controlled.

4. Packages must be clearly identified as to production lot in the event of a process deviation.

PACKAGING RECOMMENDATIONS

1. Approved packaging materials are stored and handled in a manner consistent with maintaining microbial integrity.

2. Equipment sanitation programs, employee practices, and product handling procedures must minimize product contamination during the packaging process.

3. Key factors such as product temperature, vacuum level, and modified atmosphere delivery systems, must be monitored and controlled.

4. Packages must be clearly identified as to production lot in the event of a process deviation.

5. The use of containers which have been traditionally used for the marketing of shelf stable foods is not recommended for these keep refrigerated products until such time as appropriate safeguards are in place to avoid consumer confusion and the risk of temperature abuse. This concern centers on the possibility that consumer misconception may lead to temperature abuse and a botulism hazard.

Chilled Food Processing Requirements:

1. Use only fresh, high-quality, low-microbial count ingredients.

2. Prepare food in as clean an environment as possible with routine sanitary practices in place.

3. Instruct personnel in hygiene and hygienic work practices; issue a daily change of clothing, including masks and gloves.

4. Avoid cross-contamination of foods, from raw to cooked to storage.

5. Adhere strictly to recipes, and time/temperature cooking standards, for cooked and pre-cooked product.

6. Use an accurate date-labeling system for packages.

7. Apply microbiological analysis to raw food samples, the environment, equipment and work tables.

8. Use only high quality packaging materials in conjunction with proper packaging equipment.

9. Strictly ensure correct times and temperature for chilling, refrigerated storage, and distribution.

10. Ensure correct labeling, stock rotation, and instructions for preparation.

11. Implement a HACCP monitoring program.

EZO SEMI-RIGID PACKAGING SYSTEM

Jeffrey L. De Gross
Senior Process Engineer
Packaging Development Center
Sonoco Products Company
1 North Second Street
Hartsville, SC 29550

Biography

Jeff De Gross is Senior Process Engineer in Sonoco's Packaging Development Center. Jeff's primary responsibility is the technical development of Sonoco's MAGICTOP easy-opening, retortable packaging system. Jeff joined Sonoco in May, 1987 and has been involved with the MAGICTOP project for over two years.

Prior to joining Sonoco, Jeff graduated with a Bachelor's degree in Chemistry from the University of Toledo. Jeff was employed by Owens-Illinois in May, 1976 where he began his career in package development in O-I's Plastic Beverage Technology group.

Abstract

An overview of some contemporary easy-opening, retortable, plastic packaging systems will be followed by a discussion of Sonoco's MAGICTOP^R. A simple but clever difference in its design compared to other easy-opening, heat seal systems results in MAGICTOP's very strong, abuse resistant, hermetic seal that is truly easy to open. MAGICTOP provides the food processor with the additional benefit of a wide sealing window, even when it is processed on conventional filling and sealing equipment. The combination of MAGICTOP peelable containers, conventional lower cost lidding materials, a wide sealing process window, and an abuse resistant fusion seal, all coupled with true easy opening, makes the MAGICTOP system an attractive choice for shelf stable packaging.

MAGICTOP®

THE WORLD'S FIRST EASY-OPENING SEMI-RIGID PACKAGING SYSTEM

I. PREFACE

WHAT PRICE ARE YOU PAYING FOR PRODUCT SAFETY IN SHELF STABLE CONVENIENCE FOOD PACKAGING?

THE ANSWER COULD VERY WELL BE :

- *PACKAGING FOR THE PRODUCT IS LESS "CONSUMER FRIENDLY"*

- *OVER DESIGN, LIMITING OPTIMUM PACKAGING COST*

THESE ARE EXTREMELY HIGH PRICES TO PAY. BECAUSE SAFETY IS NEVER COMPROMISED, SOME PACKAGING DECISIONS FOR THESE FOOD PRODUCTS HAVE BEEN TOUGH TO MAKE. EVEN COST DOES NOT YIELD TO SAFETY. SOMETHING HAS TO GIVE AND OFTEN IT IS USER FRIENDLINESS.

ALONG WITH THE ISSUE OF NOT COMPROMISING PRODUCT SAFETY IS THE NEED FOR WHAT THE CONSUMER'S PERCEPTION OF SAFETY IS. THIS PERCEPTION IS THEN DESIGNED INTO THE PACKAGING, RESULTING INTO HIGHER UNIT COST.

CLEARLY, THE DECISIONS TO GO TO A FULL PANEL EZO DOUBLE SEAMED END ON THE BUCKET STYLE CONTAINER AFFORDED THE FOOD PROCESSOR THE SAFETY AND CONSUMER PERCEPTION THAT WAS NEEDED TO BE SUCCESSFUL IN THE SHELF STABLE FOOD MARKET.

ON THE CONSUMER SIDE, THE BUCKET MAINTAINS THE LOOK AND PERCEPTION OF A CAN, IN WHICH THEY HAVE A GREAT DEAL OF CONFIDENCE, AND PROVIDES SOME MEASURE OF EASY OPENING QUALITY.

FOR THE FOOD PROCESSOR, THE CLOSING MACHINERY, LINE SPEEDS, AND INTEGRITY TESTING METHODOLOGY ARE FAMILIAR. THIS 'OVERALL FAMILIARITY' WITH DOUBLE SEAM CLOSING HAS RESULTED IN INCREASED SALES AND NOW, A GOOD SAFETY HISTORY.

HOWEVER, IN ORDER TO HAVE ALL OF THE ABOVE, THE FOOD PROCESSOR IS SACRIFICING A GOOD DEAL OF PROFIT DUE TO THE RELATIVELY HIGH UNIT COST OF THIS PACKAGING FORM. WHILE THERE APPEARS TO BE ROOM FOR UNIT COST REDUCTION, IT IS BEING QUICKLY FOLLOWED BY PRICE EROSION ON THE STORE SHELF.

AGAIN, WHAT IS THE PRICE PAID FOR PRODUCT SAFETY?

FOR THE BUCKET, PERHAPS IT CAN BE SUMMARIZED THAT THE PRICE IS:

• HIGHER PACKAGING UNIT COST -

- *METAL EZO ENDS HAVE HIGH UNIT COST COMPARED TO A SEMI-RIGID OPTION*

- *WITH METAL EZO, THERE IS THE ABSOLUTE NEED FOR AN OVERCAP*

- *FOR DOUBLE SEAMING, TIGHT TRIM TOLERANCES AND FLANGE THICKNESS VARIATION REDUCE LINE EFFICIENCIES.*

- *CONSUMER PERCEPTION OF SAFETY HAS LED TO THICK SIDEWALLS FOR A RIGID "LOOK AND FEEL"*

• LESS THAN OPTIMUM USER FRIENDLINESS -

- *A METAL EZO END, WHILE MANAGEABLE, IS NOT TRULY AN EASY-OPENING FEATURE FOR THE ENTIRE MARKET*

THE ASSUMPTION HERE IS THAT UNIT PACKAGING COST REDUCTION CAN ONLY GO SO FAR WITH BUCKET-TYPE PACKAGING FORM AS IT EXISTS TODAY.

IN THE AREA OF TRAY TYPE PRODUCTS, PERHAPS NO CLEAR-CUT PACKAGING CHOICE EXISTS THAT ADDRESSES THE ISSUES AS WELL AS THE METAL EZO BUCKET. EASY OPENING TRAYS WITH EZO METAL ENDS ARE NOT UTILIZED. (HORMEL'S HAM PRODUCT BEING CONSIDERED A CAN)

TYPICALLY, TRADITIONAL HEAT SEALING METHODS HAVE BEEN ADOPTED FOR CLOSING SHELF STABLE TRAY PRODUCTS. HOWEVER, THIS METHOD HAS NOT BEEN ABLE TO SATISFY TWO OPPOSING FUNCTIONS, i.e., A TIGHT SEALABILITY FOR PROTECTION OF THE PRODUCT AND CONSUMER SAFETY ON ONE HAND, AND A TRUE EASY OPENING QUALITY ON THE OTHER HAND.

WHAT HAS BEEN THE PRICE FOR PRODUCT SAFETY IN A TRAY FORM?

- •LACK OF HIGH INTEGRITY HEAT SEALS THAT ARE TRULY EASY OPENING
- • NARROW SEALING PROCESS WINDOW
- • EXTREMELY LOW LINE SPEEDS
- • NEED FOR HIGHER COST PEELABLE RETORTABLE LIDSTOCKS
- • 200% HAND INSPECTION OF SEALS
- •100% INCUBATION

THE LEAN TOWARD HIGHER INTEGRITY SEALS FOR SAFETY AND PROCESSING EFFICIENCY REASONS RESULTS IN HIGH OPENING FORCES. THIS COUPLED WITH USE OF HIGHER COST PEELABLE RETORTABLE LIDSTOCKS APPLIED AT DRAMATICALLY LOWER LINE SPEEDS COMPARED TO DOUBLE SEAMING ARE EVIDENCE THAT TRAY PRODUCT PACKAGING HAS NOT ENJOYED THE DEGREE OF OPTIMIZATION AS THAT OF THE BUCKET.

SEMI-RIGID PACKAGING OFFERS EXCELLENT POTENTIAL TO THE FOOD PROCESSOR TO BE COST EFFECTIVE, ESPECIALLY FROM A UNIT PACKAGING COST BASIS. IT OFFERS THE ABILITY TO PROVIDE TRUE USER FRIENDLINESS AT REDUCED PACKAGING COSTS AND AT A HIGH LEVEL OF PACKAGE INTEGRITY.

SONOCO BELIEVES THAT SEMI-RIGID PACKAGING WILL NEED THE FOLLOWING ATTRIBUTES TO BE CONSIDERED THE MOST COST EFFECTIVE OPTION:

- *A HIGH DEGREE OF SAFETY AND PRODUCT RELIABILITY*

  - BY HAVING A HIGH INTEGRITY, WELDED, NON-DESTRUCT HEAT SEAL

  - BY SEPARATING HEAT SEAL FROM PRODUCT OPENING FEATURE

- *A VERY CONSUMER FRIENDLY PACKAGE*

  - WITH TRUE EASY-OPENING (TYP. 4- 5# 'POP' & 1-2# 'PEEL')

- *A HIGH POTENTIAL FOR UNIT COST REDUCTION*

  - BY USING LOWER COST LIDSTOCK - UNPEELABLE AT THE HEAT SEAL- WITH
    NO SPECIAL HEAT SEAL COATINGS FOR DUAL PURPOSE
  - USE OF ALL PLASTIC LIDSTOCKS WITHOUT OVERCAP ala. DELMONTE
    VEGETABLE CLASSICS®

• A POTENTIAL FOR SYSTEM COST REDUCTION

- BY IMPROVED LINE EFFICIENCY WITH WIDE PROCESSING WINDOW OF SEALING CONDITIONS.

- BY SEALING WITHOUT REGARD FOR OPENING THE PRODUCT

- WITH HIGH INTERNAL PRESSURE RESISTANCE TO HANDLE HEADSPACE VARIATION IN THE RETORT

II. A. SONOCO'S MAGICTOP® SEMI-RIGID PACKAGING SYSTEM

SONOCO'S NEW MAGICTOP® PACKAGING SYSTEM IS A SEMI-RIGID PACKAGE FOR SHELF STABLE CONVENIENCE FOOD PRODUCTS THAT UTILIZES HIGH INTEGRITY, WELDED HEAT SEALS, YET PROVIDES A TRUE EASY OPEN FEATURE.

THE MAGICTOP SEMI-RIGID PACKAGING SYSTEM ADDRESSES THE KEY ATTRIBUTES THAT WILL LEAD TO A MORE COST EFFECTIVE PACKAGE .

FOR MAGICTOP, THE PRICE FOR SAFETY, DOES NOT INCLUDE LACK OF USER FRIENDLINESS.

THE MAGIC OF MAGICTOP PACKAGING IS SIMPLY THE USE OF A HIGH INTEGRITY, WELDED HEAT SEAL AND A COEXTRUDED SHEET WITH CONTROLLED INTERLAMINAR STRENGTH WHICH BY ITSELF DETERMINES THE OPENING FORCE. BECAUSE OF THESE TWO CENTRAL POINTS, MAGICTOP:

- HAS SEPARATE PEELING AND SEALING "ZONES".
- IS A TRUE, EASY OPENING PACKAGE
- HAS EXCELLENT RESISTANCE TO INSIDE PRESSURE
- OFFERS A WIDE SEALING WINDOW
- UTILIZES LOWER COST LIDSTOCKS WITHOUT HEAT SEAL COATINGS

A STRONG, WELDED HEAT SEAL, WITH A SEAL STRENGTH THAT EXCEEDS THE BURST PRESSURE OR INTERLAMINAR STRENGTH OF THE CONTAINER AND LID IS CONCEIVABLY THE "SAFEST" HEAT SEAL FROM THE STANDPOINT OF BIO-INTRUSION. ANYTHING LESS THAN A WELDED HEAT SEAL IS SEEMINGLY COMPROMISING THE SAFETY OF THE PRODUCT FOR OPENABILITY.

COMPETITIVE NON-WELDED LIDDING SYSTEMS, WHEREBY OPENING IS FACILITATED BY A DESTRUCT HEAT SEAL OR "COHESIVE FAILURE" IN THE SEAL ITSELF REQUIRES THE FOOD PROCESSOR TO CONCERN HIMSELF WITH THE CONSUMER'S ABILITY TO OPEN THE PACKAGE IN ADDITION TO PROVIDING A SECURE SEAL. THIS MEANS A NARROWING OF THE PROCESSING WINDOW FOR GOOD COMPROMISE BETWEEN FOOD SAFETY AND EASE OF OPENING.

THE MAGICTOP SYSTEM IS DESIGNED TO OVERCOME THE INADEQUATE POINTS OF THE CONVENTIONAL EASY PEELABLE SYSTEM BY SEPARATING THE SEALING ZONE FROM THE PEELING ZONE. THIS ALLOWS A DESIGNED-IN DIFFERENCE IN THE OPENING FORCE AND STRENGTH OF THE SEAL. WITH MAGICTOP, THE PROCESSOR ESSENTIALLY SEALS THE CONTAINER AS IF IT WERE MEANT TO BE CUT OPEN WITH A KNIFE, YET ACTUAL EASY-OPENING IS MAINTAINED.

B. THE BASIC STRUCTURE

THE MAGICTOP MECHANISM IS REALIZED BY VIRTUE OF A UNIQUE **PEELABLE SHEET**.
THIS COEXTRUDED SHEET IS THERMOFORMED INTO PLASTIC TRAYS OR OTHER SHAPES.
(SEE FIG. 1)

(FIG.1) Basic Structure of MAGICTOP

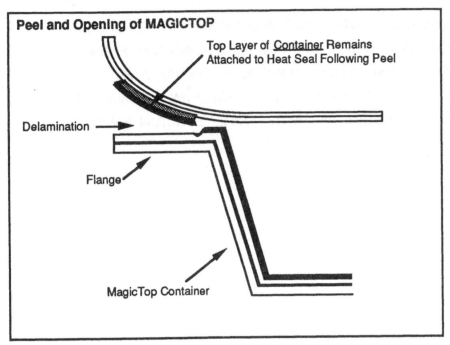

Peel and Opening of MAGICTOP

Top Layer of <u>Container</u> Remains
Attached to Heat Seal Following Peel

Delamination

Flange

MagicTop Container

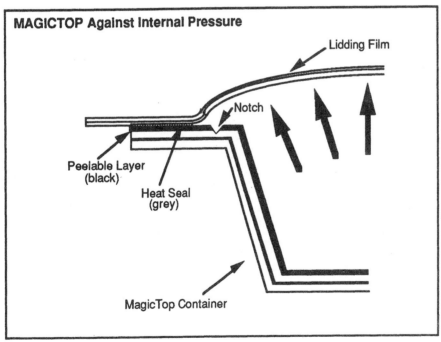

MAGICTOP Against Internal Pressure

Lidding Film

Notch

Peelable Layer
(black)

Heat Seal
(grey)

MagicTop Container

THE INTERLAMINAR STRENGTH BETWEEN THE INNERMOST LAYER (PEELING LAYER) AND THE SECOND LAYER IS CONTROLLED AND TARGETED AT ~2.5 - 3.0 LBS/IN. THE SECOND KEY TO THE MECHANISM IS THE MECHANICAL PLACEMENT OF A "NOTCH" ONTO THE INNER FLANGE OF THE CONTAINER. (ALSO SHOWN IN FIG.1)

A NON-PEELABLE LIDDING MATERIAL IS WELD-SEALED TO THE INNERMOST LAYER OF THE CONTAINER OUTSIDE THE NOTCH. UPON OPENING, THE PEELING LAYER OF **THE CONTAINER** IS REMOVED SIMULTANEOUSLY WITH THE LID AND DELAMINATES THE CONTAINER UP TO THE NOTCH. A 'RING' OF PEELING LAYER MATERIAL FROM THE SHEET IS REMOVED ENTIRELY, REMAINING ATTACHED TO THE LID WITH HEAT SEAL INTACT.

CONVERSELY, IT IS NOT SO EASY TO START DELAMINATION OF THE PEELING LAYER FROM THE INSIDE. IN FACT, MAGIC TOP IS DESIGNED SUCH THAT THE FORCE TO DELAMINATE THE PEELING LAYER FROM THE NOTCH OUTWARD USUALLY EXCEEDS EITHER THE INTERLAMINAR STRENGTH OF LIDSTOCK LAYERS OR ITS BREAKING STRENGTH.

(FIG.2) Different in Strength of MAGICTOP Containers

Lidding Film

Stress Concentration

**Opening from Outside
Typically.
4 to 5 lbs "Pop"
1 to 2 lbs "Peel"**

**Opening from Inside
Typically >10 lbs/in**

TO OPEN MAGICTOP, THE SURFACE LAYER OF THE CONTAINER BODY IS EASILY PEELED BECAUSE OPENING FORCE IS CONCENTRATED PRECISELY AT THE HEAT SEAL AND THE INTERFACE OF THE PEELING LAYER AND 2ND LAYER (SEE FIG. 2, LEFT FIGURE) THE PEELING STRENGTH DURING OPENING IS ONLY DEPENDENT UPON THE SHEET PEEL STRENGTH (DEFINED AS THE INTERLAMINAR STRENGTH OF THE SHEET'S INNERMOST LAYER TO THE SECOND LAYER OF THE SHEET). IT IS LARGELY INDEPENDENT ON SEALING CONDITION AS WILL BE SHOWN.

THE OPENING FORCE FROM THE INSIDE OF THE MAGICTOP PACKAGE IS A QUITE DIFFERENT PHENOMENON. AGAIN, FORCES ARE CONCENTRATED AT THE HEAT SEAL, BUT AWAY FROM THE NOTCH AS SHOWN BY ARROW IN FIG.2. BY CONCENTRATING FORCE AWAY FROM THE INTERFACE OF THE TWO LAYERS, ANOTHER FACTOR NOW CONTROLS THE OPENING FORCE BESIDES INTERFACIAL STRENGTH OF LAYERS 1 AND 2.

(FIG.3) Opening from Inside

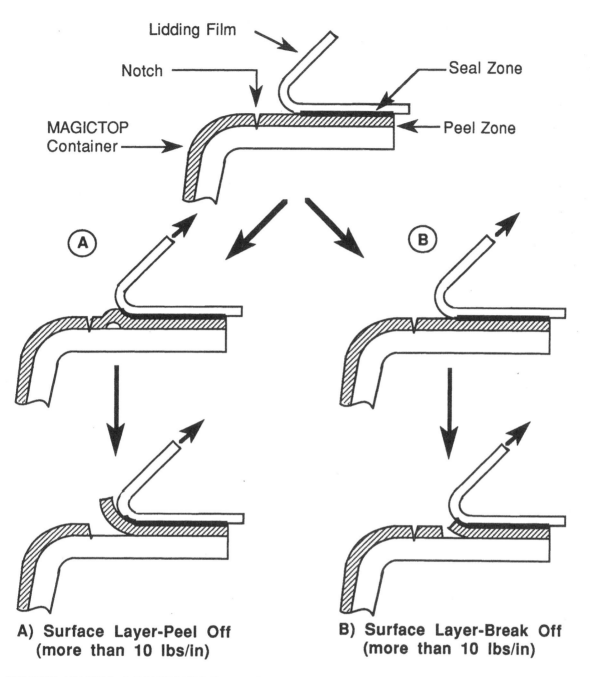

Lidding Film

Notch

Seal Zone

MAGICTOP
Container

Peel Zone

Ⓐ

Ⓑ

**A) Surface Layer-Peel Off
(more than 10 lbs/in)**

**B) Surface Layer-Break Off
(more than 10 lbs/in)**

TO EXPLAIN THIS, A SCHEMATIC REPRESENTATION FOR INSIDE FAILURE IS SHOWN IN FIG.3.
TWO FAILURE MODES ARE SHOWN. TO THE LEFT IS AN INSIDE DELAMINATION OF THE
PEELING LAYER FROM THE REMAINDER OF THE CONTAINER BODY AND AT THE RIGHT AN
ACTUAL BREAK IN THE PEELING LAYER BETWEEN THE HEAT SEAL AND THE NOTCH.

IT IS CLEAR THAT THE CONTROLLING FACTORS FOR THE INSIDE FAILURE TYPES SHOWN HERE ARE : 1) THICKNESS OF SHEET PEEL LAYER AND 2) THE BOND STRENGTH BETWEEN PEEL LAYER AND SECOND LAYER OF SHEET. BECAUSE THE SEALING STRENGTH IS A COMBINATION OF THE PHYSICAL STRENGTH OF THE SHEET PEELING LAYER AND THE INTERLAMINAR STRENGTH BETWEEN THE PEELING LAYER AND THE SHEET SECOND LAYER, IT IS POSSIBLE TO ADJUST AS REQUIRED, THE "SEALING STRENGTH" OF MAGICTOP BY CHANGING EITHER OF THESE TWO PARAMETERS.

C. NOTCHING

BECAUSE OF THE PRINCIPLE PREVIOUSLY EXPLAINED, IT WAS EXTREMELY IMPORTANT TO ESTABLISH A RELIABLE NOTCHING TECHNOLOGY. SONOCO BENEFITS FROM THE NEARLY FOUR YEARS OF PRODUCTION EXPERIENCE OF OUR LICENSOR, IDEMITSU PETROCHEMICAL, WHO BY INTENSIVE RESEARCH AND DEVELOPMENT HAS INVENTED A RELIABLE PROCESS THAT ALLOWS US TO OFFER NOTCHED, COEXTRUDED MAGICTOP CONTAINERS.

OBVIOUSLY, CONTAINERS ARE SUPPLIED ALREADY NOTCHED TO THE FOOD PROCESSOR. WHILE NOTCHING OF PLASTIC CONTAINERS TO A HIGH DEGREE OF PRECISION IS SEEMINGLY DIFFICULT, PERHAPS IT CAN BE PUT INTO PERSPECTIVE THAT TARGET NOTCH DEPTHS ARE DESIGNED TO ENCOMPASS THE STACKUP IN TOLERANCES THAT OCCUR DUE TO INDIVIDUAL LAYER THICKNESS VARIATION IN THE CONTAINER AND THE FLANGE THICKNESS VARIATION.

GENERALLY SPEAKING, THE PEELING LAYER THICKNESS IS ON THE ORDER OF 10% OF THE TOTAL SHEET THICKNESS. FOR A 50 MIL FLANGE, THIS RESULTS IN A PEEL LAYER THICKNESS OF APPROXIMATELY 5 MILS +/- 15% FOR LAYER NON-UNIFORMITY REPRESENTING A TOTAL RANGE OF PERHAPS 4-6 MILS.

THE ONLY REQUIREMENT FOR NOTCH DEPTH IS TO ASSURE THAT THE PEELING LAYER IS COMPLETELY CUT THROUGH. OUR NOTCH DEPTH TARGET FOR THE 50 MIL FLANGE EXAMPLE IS NEARLY DOUBLE THE IDEAL DEPTH, ABOUT .010", FULLY ENCOMPASSING THE RANGE AND STACKUP OF TOLERANCES. TYPICAL DEPTH VARIATION IS +/- 2 MILS.

III. PERFORMANCE OF MAGICTOP

A. A WIDE PROCESSING WINDOW

MAGICTOP CONTAINERS ARE SEALED BY CONVENTIONAL CONDUCTION HEAT SEALING
METHODS. AS WITH ANY HEAT SEAL, TEMPERATURE, PRESSURE AND TIME (DURATION OF
THE SEAL BAR) AFFECT THE SEAL INTEGRITY.
FOR MAGICTOP THE OPERATING WINDOW IS BELIEVED TO BE INCREASED.

TO PROVIDE SOME INDICATION AS TO HOW WIDE THE PROCESSING WINDOW IS FOR
MAGICTOP, A SIMPLE EXPERIMENT WAS DEVISED. IN THIS EXPERIMENT, MAGICTOP
CONTAINERS WERE HEAT SEALED AT A VARIETY OF SEAL BAR TEMPERATURES BEGINNING
AT 170°C, INCREASING THE TEMPERATURE BY 5°C UP TO 210°C. PRESSURES AND DWELL
TIMES WERE KEPT CONSTANT.

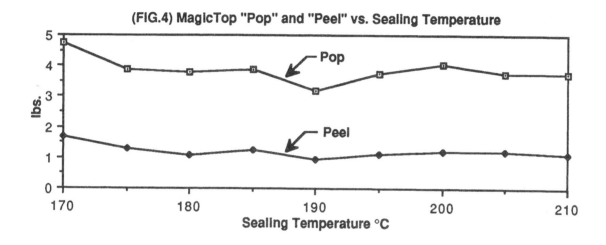

(FIG.4) MagicTop "Pop" and "Peel" vs. Sealing Temperature

WHOLE PACKAGE "POP & PEEL" TEST DATA WERE TAKEN ON CONTAINERS PRODUCED AT
EACH OF THE TEST TEMPERATURES. AS SEEN IN A GRAPH OF THIS DATA (SEE FIG.4), PEEL
STRENGTH FOR MAGICTOP REMAINS ESSENTIALLY THE SAME THROUGHOUT THE
TEMPERATURE RANGE, AROUND 1.5 LBS/IN, INDICATING A VERY WIDE RANGE AT WHICH
MAGICTOP CAN BE SEALED WITHOUT AN AFFECT ON THE PEELING FORCE. 'POP' FORCE,
DEFINED AS THE INITIAL OPENING FORCE, HAS A SIMILAR RESULT. 'POP' FORCE REMAINS
QUITE FLAT AT AN AVERAGE OF AROUND 4-5 LBS/IN ACROSS THE RANGE OF SEALING
TEMPERATURES.

FOR COMPARISON, TWO COMMERCIALLY AVAILABLE PEELABLE LIDSTOCKS WERE
SUBJECTED TO IDENTICAL SEALING CONDITIONS UTILIZING NON-MAGIC TOP PP CONTAINERS
(GEOMETRY IDENTICAL). PEEL STRENGTHS FOR THESE WERE MEASURED AND PLOTTED. IN
FIG 5, THE RESULTS FOR THE COMMERCIALLY AVAILABLE RETORT LIDSTOCKS APPEAR
GRAPHICALLY ALONG WITH THE MAGICTOP DATA. NOTE THAT THE PEEL STRENGTH IS ~.25
LBS/IN AT 170°C AND IS STILL <2 LBS/IN AT 190°C. HOWEVER BOTH COMMERCIAL LIDSTOCKS
BECOME NON OPENING AT ANY TEMPERATURE ABOVE 190°C INDICATING NOT AS BROAD A
SEALING TEMPERATURE WINDOW AS MAGICTOP.

(FIG.5) Peel Strength vs. Sealing Temperature

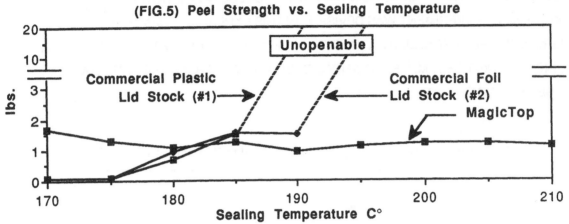

THESE RESULTS SHOW THAT, ASSUMING GOOD HEAT SEALS, MAGICTOP PACKAGING IS
TRULY AN EASY OPENING SYSTEM COMPARED WITH COMPETITIVE EASY PEEL LIDSTOCKS.

IN ORDER TO LEND SUPPORT THAT THE MAGICTOP CONTAINERS USED IN THIS EXPERIMENT
EXHIBITED A WORTHY HEAT SEAL, FIG. 6 SHOWS THE RETRAINED BURST RESULTS AT EACH
OF THE SEAL TEMPERATURES FOR MAGICTOP AS WELL AS THE COMMERCIAL LIDSTOCKS.
NOTE THAT THROUGHOUT THE TEMPERATURE RANGE, VERY HIGH BURST PRESSURES
WERE OBTAINED FOR MAGICTOP. IN ADDITION, THE BURST TEST RESULTS REMAINED
FAIRLY LEVEL THROUGH THE ENTIRE 40°C (104°F) TEMPERATURE RANGE - CLEARLY AN
INDICATION OF A WIDE PROCESSING WINDOW.

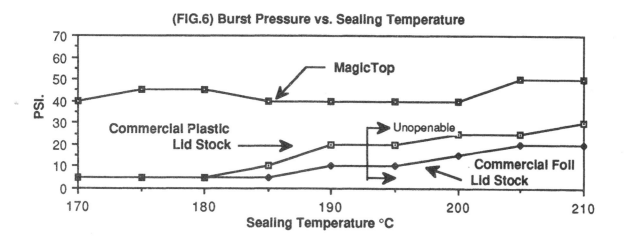

THE COMMERCIAL LIDSTOCKS DID NOT APPEAR TO HAVE A STRONG HEAT SEAL UNTIL VERY
NEAR THE TEMPERATURE AT WHICH IT WAS DETERMINED NON-OPENING IN THE PEEL TEST
AS EVIDENCED BY THEIR BURST RESULTS.
THIS EXPERIMENT WAS NOT INTENDED TO REPRESENT IDEAL SEAL CONDITIONS FOR THE
CONVENTIONAL LIDSTOCKS WITH COHESIVE FAILURE MECHANISM. HOWEVER, THERE IS
CLEAR INDICATION THAT THE TEMPERATURE PROCESSING WINDOW, FROM THE
STANDPOINT OF PEELING STRENGTH AND RESISTANCE TO BURST, IS INCREASED FOR
MAGICTOP OVER COMPETITIVE SYSTEMS.

B. PEEL PERFORMANCE
COMPARISON OF COMMERCIAL <u>OFF THE SHELF, AFTER RETORT CONTAINERS WITH</u> <u>MAGICTOP</u>

WE HAVE SEEN SO FAR A WIDE PROCESSING WINDOW WITH REGARD TO SEALING MAGICTOP PACKAGING. PEEL STRENGTHS AND BURST TEST RESULTS ARE ESSENTIALLY UNCHANGED ACROSS A WIDE RANGE OF TEMPERATURES. THROUGHOUT THIS RANGE WE HAVE SHOWN THAT THE SEAL INTEGRITY IS EXTREMELY HIGH AS MEASURED BY BURST TEST RESULTS. FURTHER, WE HAVE SHOWN THAT MAGICTOP PACKAGING YIELDS A TRUE EASY OPENING AT A WIDE RANGE OF SEALING TEMPERATURES. ALL THIS WAS PRODUCED UNDER CONTROLLED LABORATORY CONDITIONS.

A SIDE-BY-SIDE PEEL/SEAL COMPARISON WAS MADE OF COMMERCIAL, RETORTED MAGICTOP CONTAINERS AND EXISTING U.S. SHELF STABLE CONTAINERS UTILIZING CONVENTIONAL PEELABLE LIDDING FILMS.

FIGURE 7 SHOWS A DIRECT COMPARISON OF THE INITIAL 'POP' FORCE TO INITIATE OPENING, THE CONTINUOUS PEELING FORCE AND THE BURST PRESSURE OF A VARIETY OF COMPETITIVE CONTAINERS AND MAGICTOP. ALL CONTAINERS WERE REMOVED FROM THE STORE SHELF, INCLUDING THE MAGICTOP EXAMPLES, WITH NO SPECIAL PREPARATION.

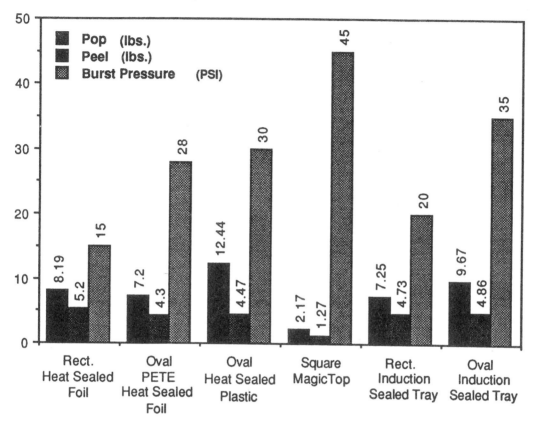

(FIG.7) Commercial Container Test

THREE POINTS ARE NOTEWORTHY:

FIRST, COMMERCIAL MAGICTOP NOT ONLY HAS THE LOWEST AFTER-RETORT POP AND PEEL STRENGTH IN A SIDE-BY-SIDE MEASUREMENT WITH COMPETITION PACKAGING, IT ALSO HAS THE HIGHEST BURST PRESSURE OF ANY OF THE EXAMPLES. SECOND, IN ALL CASES, THE CONTINUOUS PEEL STRENGTH FOR THE COMPETITOR CONTAINERS ARE ALL MORE THAN TWICE THE *POP FORCE* OF THE COMMERCIAL MAGIC TOP. THIRD, CLEARLY PRODUCT SAFETY IS DEMONSTRATED BY THE EXCELLENT BURST TEST RESULTS. FOR ALL CONTAINERS, SAFETY IS NOT COMPROMISED, BUT.........

WHAT IS THE PRICE FOR SAFETY? DOES THE PACKAGING PRICE LIST HAVE TO INCLUDE A HIGH DEGREE OF PRODUCT SAFETY AND RELIABILITY?.. THE ABILITY TO RUN AT HIGHER SPEEDS IN A ROBUST FILLING AND SEALING ENVIRONMENT?.. THE ABILITY FOR THE CONTAINER TO ACCOMMODATE A VARIATION IN HEADSPACE AND STILL SURVIVE THE HOSTILE ABUSE OF YOUR RETORT PROCESS?.. THE ABILITY FOR THE CONSUMER TO HAVE THE CONVENIENCE THEY DESIRE IN AN OPENING FEATURE?

WE BELIEVE MAGICTOP CAN OFFER YOU THIS WITHOUT PAYING THE PRICE.

THE MAGICTOP SEMI-RIGID PACKAGE OFFERS A HIGH INTEGRITY, WELDED HEAT SEAL WITHOUT:

- TRADEOFF IN OPENING FORCE

- NARROW SEALING PROCESS WINDOW

- POTENTIAL FOR BAD SEALS AND SLOWER SEALING SPEEDS DUE TO "DUAL PURPOSE-PROCESSING"

- THE ADDITIONAL COST BUILT INTO LIDSTOCKS WITH PEELABLE HEAT SEAL COATINGS

HIGH INTEGRITY, WELDED HEAT SEALS SUCH AS MAGICTOP'S HAVE THE POTENTIAL OF:

•HIGHER LINE SPEEDS. FILL/SEAL SYSTEMS USED IN EUROPE FOR 'READY MEALS' HAVE EXCELLENT POTENTIAL TO PROVIDE COST EFFECTIVE PROCESSING WITH EXISTING TECHNOLOGY, GIVEN THE RIGHT PACKAGING.

•AN IMMEDIATE 'FIT' WITH TODAY'S HIGHEST SPEED SEALERS. READY MEALS IN EUROPE ARE SEALED TO PROVIDE NON-DESTRUCT HEAT SEALS -> LIKE MAGICTOP'S.

• NON-DESTRUCT SEALS AND PROCESSING METHODS MAY ALLOW YOU TO MORE QUICKLY DEMONSTRATE TO THE USDA THAT 100% INCUBATION WILL NOT BE REQUIRED IN YOUR PLANT.

IN CONCLUSION, MAGICTOP CAN OFFER YOU:

A HIGH DEGREE OF CONFIDENCE IN PRODUCT SAFETY -

- OBTAINED BY MAGICTOP'S HIGH INTEGRITY, WELDED, NON-DESTRUCT HEAT SEALS.

TRUE EASY OPENING

- 4-5 LBS 'POP' AND 1-2 LBS CONTINUOUS PEEL

IMPROVED PROCESSABILITY

- NO "DUAL-PURPOSE" PROCESSING OR MATERIALS ARE REQUIRED

THE POTENTIAL FOR REDUCED SYSTEM COST -

- WITH WIDER SEALING PROCESS WINDOW
- LESS "BAD SEALS" YIELD HIGHER EFFICIENCIES
- HIGH INTERNAL PRESSURE RESISTANCE ALLOWS HEADSPACE VARIATION

THE POTENTIAL FOR REDUCED UNIT PACKAGING COST

- WITH LOWER COST LIDSTOCKS.

WE CHALLENGE THE INDUSTRY TO ALLOW SONOCO TO SHOW THAT THE MAGICTOP SYSTEM HAS THE BEST POTENTIAL TO OFFER THE LOWEST COST, A TRULY FUNCTIONAL SEMI-RIGID OPTION.

Appendix I -Commercial MAGICTOP Samples:

The commercial MAGICTOP example described in this paper is a Retorted Pet Food product from Japan called "Festa". Samples of Festa were obtained by Sonoco on an overseas trip.

Appendix II - Test Container Information:

For all Test Containers described in figs. 4, 5, 6, 7 (incl. non-MAGICTOP)

MAGICTOP Container Geometry:	BUCKET-STYLE
non-MAGICTOP Container Geometry:	BUCKET STYLE
(Both produced by Sonoco)	

Gram Weight:	12 gm
MAGICTOP Lidstock:	HIGH BARRIER ALL-PLASTIC (NON-PEELABLE)
Commercial Lidstock #1:	HIGH BARRIER ALL-PLASTIC W/PEELABLE HS COATING
Commercial Lidstock #2:	PLASTIC/FOIL/PLASTIC w/PEELABLE HS LAYER

Appendix II - *EXPERIMENTAL PROCEDURE FOR POP AND PEEL:*

CONTAINERS ARE RESTRAINED IN A TEST STAND ON AN INSTRON MACHINE AT A 45° ANGLE. THE OPENING TAB IS CLAMPED BY THE UPPER INSTRON JAW, THE SLACK IS TAKEN UP AND THE TEST IS STARTED. FOR THE INITIAL 'POP', CROSSHEAD SPEED IS X IN/MIN. CROSSHEAD SPEED IS PROGRAMMED TO CHANGE TO Y IN/MIN IMMEDIATELY AFTER POP TO MORE CLOSELY RESEMBLE THE SPEED OF DELAMINATION DURING AN ACTUAL PEEL. THE HIGHEST FORCE TO INITIATE PEEL IS RECORDED AS THE 'POP' FORCE. THE PEEL IS ACTUALLY AN AVERAGE OF THE LOAD CELL READINGS RECORDED FROM 'POP' UNTIL THE FORCES BEGIN CLIMBING AGAIN AT 'PULL OFF' OF THE LID. PULL OFF IS PARTICULARLY GEOMETRY DEPENDENT AND WAS THEREFORE NOT INCLUDED IN WHAT IS TERMED THE PEEL STRENGTH FOR ANY OF THE TEST CONTAINERS.

APPENDIX III -*EXPERIMENTAL PROCEDURE FOR BURST TEST*

IN OUR RESTRAINED BURST TEST, EXAMPLE CONTAINERS ARE PLACED IN A STAND BETWEEN TWO LEXAN PLATES. THE TOP PLATE IS ADJUSTED 1/16" ABOVE THE CONTAINER FLANGE. A SEPTUM IS PLACED IN THE CENTER OF THE LID FORMING A SEAL ON THE UNDERSIDE OF THE TOP PLATE OF THE BURST TEST APPARATUS.

COMPRESSED AIR IS BROUGHT IN BY MEANS OF A GC-TYPE HEADSPACE SYRINGE. INITIALLY, THE CONTAINER IS PRESSURIZED TO 15 PSI. THIS PUSHES THE LIDDING MATERIAL UP AROUND THE SEPTUM AND INTO INTIMATE CONTACT WITH THE UNDERSIDE OF THE THE LEXAN TOP PLATE OF THE TEST STAND. ONCE THE TECHNICIAN IS SATISFIED THERE IS NO LEAKAGE AROUND THE SEPTUM, THE PRESSURE IS INCREASED TO 20 PSI. IT IS HELD AT THIS PRESSURE FOR 30 SEC. AFTER THE 30 SEC TIME PERIOD, THE PRESSURE IS INCREASED 5 PSI AND HELD FOR ANOTHER 30 SEC, FOLLOWED BY INCREMENTAL INCREASE OF 5 PSI AND 30 SEC HOLD TIME PERIODS. WE GENERALLY CONSIDER REACHING THE 30PSI LEVEL AND HOLD FOR 30 SEC A "PASS" CONDITION.

INTERACTION OF PACKAGING & FOODS TO PROVIDE
SUPERIOR QUALITY MICROWAVEABLE FOOD PRODUCTS

Robert W. Fisher
Senior Program Manager
Campbell Institute for Research & Technology
Campbell Soup Company
Campbell Place
Camden, NJ 08103-1700

Biography

Bob is a native of New Jersey and received his Ph.D. in Food Science from Rutgers - The State University of NJ in 1982. Since that time he has been employed by the Campbell Soup Company, Campbell Institute for Research and Technology in Camden, NJ. During the past eight years he has been in Product Technology, Beverage Product Development, Frozen Food Product Development and Microwave Technology. Currently, he is Senior Program Manager for Traditional Meals, Swanson Dinners Product Development, Microwave Technology and the Campbell Microwave Institute.

Bob is a member of the Institute of Food Technologists, the International Microwave Power Institute, Sigma Xi, Phi Tau Sigma, the Product Development Management Association, American Management Association and the Executive Advisory Panel for Food Engineering.

Abstract

In the design of microwaveable food products the interaction of microwave energy, food and package are critical to the development of a successful product.

A "Total Systems" approach to product design will provide opportunities to develop superior quality microwaveable products in the future. Developments in packaging, oven design and other technologies will be discussed.

"INTERACTION OF MICROWAVE PACKAGING AND

FOOD FORMULATION TO PROVIDE SUPERIOR QUALITY

FOOD PRODUCTS"

DR. ROBERT W. FISHER
SENIOR PROGRAM MANAGER
CAMPBELL MICROWAVE INSTITUTE
CAMPBELL SOUP COMPANY
CAMDEN, N.J.

PRESENTED AT

FOODPLAS '91

PLASTICS INSTITUTE OF AMERICA
ORLANDO AIRPORT MARRIOTT HOTEL

ORLANDO, FLORIDA
MARCH 5 - 7, 1991

Today we are going to take a brief look at the interaction of microwave packaging and food formulation to provide superior quality food products.

To begin, lets discuss the current status of the microwave oven in the world today. The microwave oven continues to provide a unique opportunity and challenge to the food industry for new product and package development. Since microwave penetration exceeds 80% of households and is present in 75% of the workplaces, in America, there is and will continue to be, a growing need to develop packages and products which are "designed" for the microwave. The current microwave product marketplace is estimated to be approximately $2.3 billion dollars in retail sales, which includes foods adapted and targeted for microwave use. This market has increased in size 9.7% versus a year ago. Over the past seven to eight years the food industry and packaging manufacturers have made a significant commitment to the development of microwavable products. This has been driven by microwave penetration and consumer needs as a result of changes in lifestyle, demographics and technological developments.

While the microwave revolution was led by America, it is truly a global phenomenon. The U.S. leads worldwide penetration with approximately 83% of the homes having one oven, while more than half of the homes in Japan and Australia have microwaves. Europe follows, with some countries having relatively low penetration rates but high growth rates as illustrated in Table 1.

At Campbell's, we are concerned about the microwavability of the product. We see microwavable products as "Total Systems" and work to design, develop and provide quality value-added microwavable products for our consumers. The "Total System" includes; formulation, package design, the microwave oven and energy along with heating directions. To design these products we must understand the interaction of the microwave energy, food and package. The finished product must act as a total system in the microwave during heating to provide consumers with the quality product they expect in a convenient manner.

MICROWAVE ENERGY FUNDAMENTALS

Before we can design a product, we must understand how microwave energy interacts with the package and the food. First, let's discuss what microwave energy is and how it interacts with different materials. Microwaves are electromagnetic energy which are similar, in nature, to visible light, x-rays, radio waves (UHF, VHF) and ultraviolet energy. All electromagnetic waves are composed of rapidly alternating electric and magnetic fields which oscillate at different rates. The energy frequency, in domestic retail ovens is 2,450 MHz, while industrial ovens commonly used in plant manufacturing operations operate at 915 MHz. This means that the field in household ovens is alternating at 2,450 million times per second. Heating begins to occur when charged compounds, such as water, try to align within this rapidly changing field. Water, which is present in foods at approximately 50 - 95%, contains positively charged hydrogen

352

atoms and negatively charged oxygen atoms which cause the water molecule to move within the field. This movement causes friction amongst adjacent molecules which generates the heat. Ionic materials, such as salt or sodium chloride, will intensify the heating if they are in solution. Once the heat is generated it is then transferred to other points in the product by conduction and convection. The specific heat, thermal conductivity, density and viscosity all effect the rate of heat transfer.

The interaction of microwave energy with a substance begins when microwaves strike the surface of a material and are either transmitted, reflected or absorbed. When absorption occurs, penetration depth begins to play an important role. For example, if a food is surrounded by a paper towel in the microwave oven, the paper towel absorbs very little of the energy while the food product absorbs the majority of the energy and begins to heat. However, if this same product were covered with aluminum foil most of the energy would be reflected back into the oven cavity with the food receiving very little energy.

Another property of a material which is critical to microwave heating is its dielectric property. The dielectric property of a material is the physical description of how well a material can potentially heat when it interacts with electromagnetic energy. Most glass, plastic and paper packaging materials have low dielectric constants compared to food and are transparent to microwave energy. Figure 1, illustrates the penetration depth of a variety of packaging materials in comparison to, the strong absorption of microwave

353

energy by water or salt solutions. Please note that ice is transparent to microwave energy along with most packaging materials. Many factors influence the dielectric properties of food. A few of these are; moisture content, temperature, salt content, physical state and chemical composition. An example of taking advantage of these properties in product design would be to develop a hot fudge ice cream sundae which is microwavable.

MICROWAVE PROPERTIES OF FOODS

Now that we understand what microwave energy is and how it interacts with packaging and food materials, let's discuss how it can be utilized for microwavable foods. The type of food product which one must design and its physical state are key factors to determine and understand prior to initiating development of a microwavable food product. For example, whether the finished product is frozen or self stable has a tremendous impact on the complexity of the finished product with regard to its microwave properties and characteristics. The water in a frozen meal must go through a phase change, ice to water, during heating which often generates "hot spots" due to the large differential in energy absorption between water and ice. A shelf stable meal or a refrigerated product which contains water in a liquid state and does not require a phase change, could result in more uniform product heating in the microwave.

The complexity of the food system is another issue with which the food technologist and packaging engineer must deal. Complexity, in this case, is defined as the formulation of the product as influenced by heating by microwave energy rather than conventional heating. There are different levels of complexity in the designs of food formulation. These levels include; 1) ingredient interaction within a component, for example, the heating of a starch within a beef gravy, 2) the type of component and its shape, size, location and formulation, an example would be the beef gravy over a beef patty and 3) the total weight of the product and the manner in which the product is assembled. A frozen meal which contains mashed potatoes, carrots and a beef patty with gravy, is an example of a typical finished product which a consumer would purchase containing all of these interactions. Consequently, ingredient and component interaction are critical in product design. This issue can be further illustrated by the heating of the following microwavable frozen products; macaroni and cheese meat lasagna or pizza all differing dramatically in design and microwave properties.

Food products in the marketplace are virtually every shape (ellipse, square, rectangle, sphere-like), size and weight along with being in all types of packaging shapes and materials such as; aluminum, plastic, glass, paper and susceptors. The determination of food size, shape, weight and packaging characteristics and materials have a major effect on total product design and finished product quality.

It is known that uneven heating can be due to: the wave patterns that create "hot spots" to a varying degrees in all microwave ovens or the characteristics of the components within the product and how they absorb microwave energy (O'Meara & Reilly, 1986). The data available for these effects is very limited in the literature. What is available is usually product or ingredient specific which might apply to your situation. However, an example of some fundamental data that exists on food shape and surface area was developed by Burnett, 1990 (1). Dr. Burnett showed that when cylindrical and spherical potato samples were reheated in a microwave according to internal temperature and weight loss, both large and small volumes exhibited similar trends. The spherical samples had a significantly higher internal temperature and consequently a greater weight loss than the cylindrical samples under the same conditions as shown in Tables 2 & 3. This data shows the importance of shape and size.

Another fundamental study was conducted by O'Meara and Reilly, 1986 (6), where they determined the differential in the final temperature of different components in a model frozen dinner. The frozen dinner consisted of a meat patty, mashed potatoes and sliced carrots all placed on a round plastic tray. The dinners were heated in three different microwave ovens to obtain an internal temperature of 140° F in the meat patty. Results indicated that the heating patterns of the microwave ovens had only a small effect on the finished average temperature of the components and the relative heating rates of the components were not greatly influenced by the

orientation of the dinner in the oven. In all but one
instance, the meat and carrots were the warmest with the
mashed potatoes being appreciably colder. These results were
consistent whether heating under full power or 50% power.
Severe examples of this, were temperatures in the coldest
component measured at 30° F and 184° F one inch apart. While
temperatures of 77° F and 196° F were measured in the warmest
component and as in Figure 2.

As a result, to enable consumers to obtain optimum
quality frozen products we must conduct research in the area
of dielectric heating of foods to understand and "design in"
known factors using shape, size, weight and characteristics of
ingredients.

MICROWAVE PACKAGING CONSIDERATIONS

The driving force for the development of microwavable
packaging materials has been consumer demand. The features
these packages provide are convenience, safety, aesthetics,
enhanced food quality and a balanced price/value.
Classically, the packaging industry has been concerned with
designing packages which: protect the quality attributes of
foods, minimize the effects of physical abuse to the product
during manufacture and distribution, are relatively easy to
manufacture, economical to the consumer and simple to open and
use. With the advent of the microwave the opportunity has
availed itself for packaging technology to have a direct
effect on the design of the product and its heating
performance in the microwave oven. Microwave interactive
packaging materials such as susceptors, microwave transparent

materials (plastics, paperboard) and reflective materials
(metals) all provide both opportunities and problems for food
manufacturers to design unique new products. The selection of
the correct type of packaging is critical. This decision
should include considerations of the preferred package shape,
size, and profile along with the type of material, if the
product is to be dual ovenable or microwave only.

Alcan International Ltd. has developed the concept of
orienting microwave energy and focusing/modifying fields with
packaging to provide: more uniform heating, localized heating,
and the potential for product browning (1,2,3). Alcans'
method uses an aluminum tray married with a "MicroMatch" lid
which is designed for specific product requirements. Foods in
which this system has been utilized are pot pies, rice dishes,
potato dishes, and frozen dinners. The use of this type of
concept will be very powerful as we design both package and
product for the microwave oven.

What types of things will the future bring with regard to
microwave packaging ? Doneness indicators will enable
consumers to heat foods to a suitable temperature for
consumption without complicated instructions or the need to
remove the product from the oven. The Japanese have recently
developed a whistling doneness indicator where steam passes
through a whistling device as the food approaches the correct
temperature (7). Previous doneness indicators that have been
developed are the "3M MonitorMark" and classical pop-up
indicators. This area is ripe for development and
commercialization for microwavable products. The other factor
which needs to be mentioned is the environmental implications

of microwavable packaging. As microwavable packaging is developed in the future, we will most likely have to deal with potential recycling conflicts due to the types of materials which are used to direct and control microwave energy. For example, combinations of metals and plastics can alter microwave fields but will increase the difficulty of recycling.

MICROWAVE OVENS AND PERFORMANCE TESTING

The microwave oven is another variable we must deal with in designing products. During the development of the food product its necessary to determine its heating behavior in multiple microwaves. This is necessary since the following factors vary with ovens, such as: power level in watts, size of the cavity, type and location of the microwave energy input, and cavity wall construction.

Performance testing of the product in the microwave should be done in a minimum of three different ovens as described by Schiffman, 1987 (8) to account for fundamental differences in commercial ovens and their heating parameters.

What is in the global microwave oven market now, and what is coming in the future? The big changes are coming from the technological developments in the electronics industry. The further development of computer chips with increased speed and memory will enable manufacturers to increase internal processing complexity but provide simple external controls for the consumer. For example, a one touch key pad for heating vegetables might weigh the product, sense moisture and temperature during heating, then feedback to the master control the current state of the product and shutdown the oven when the food is heated to 140° F. This technology is available today and is present in some of our ovens. Simplicity for consumers will be continually driven by the manufacturers. Emphasis will be on simplifying the control pad making it less cluttered and utilizing symbols rather than words.

"Inverter technology" was introduced a few months ago to our shores. Sharp has patented "Inverter Technology" which has removed the transformer from the oven and replaced it with an inverter. The inverter cuts the microwave oven weight in half and provides the ability to program the microwave to operate at variable wattages ie., 525 watts or 450 or 675 watts for the time required, compared to the current ovens and their fixed wattage ie., 700 watts. This technology is inexpensive too. The Sharp inverter microwave is only $189.

CONCLUSION

In the future, as more information regarding the dielectric properties of foods becomes available, the challenge for the food technologist and packaging engineer will be to understand the the fundamental properties of microwave energy, dielectric properties of foods and packages, along with oven characteristics. These tools, if used correctly, will enable food manufactures to design products which will fulfill consumer needs on a reproducible basis providing continuing value to our customers.

REFERENCES

1. Ball, M.D., Hewitt, B., Keefer, R.M. 1985a. The Effect of Container Power Distribution on the Heating of Foods in the Microwave Oven. Presentation at the 20th Annual International Microwave Power Symposium, Chicago, IL.

2. Ball, M.D., Hewitt, B., Keefer, R.M. 1985b. Selective Heating of Multiple Food Component Meals in the Microwave Oven. Presentation at the 20th Annual International Microwave Power Symposium, Chicago, IL.

3. Ball, M.D., Hewitt, B., Keefer, R.M. 1985c. Food Browning Induced by the Intense-Field Dielectric Heating. Presentation at the 20th Annual International Microwave Power Symposium, Chicago, IL.

4. Burnett, S.A. 1990. Problems Associated with Increased Use of Microwave Energy in Food. Technical Memorandum No. 588. Campden Food & Drink Research Association.Chipping Campden, Gloucestershire, GL55 6LD, U.K.

5. Druin, M.L. 1989. Microwave Packaging at Campbell Soup Co. Presentation at The 6th Annual Eastern Food Science and Technology Conference, Hershey, Pa. October 1989.

6. O' Meara, J.P. and Reilly, D.B. 1986. The Effect of Oven Parameters On Microwave Reheating of Frozen Dinners. Microwave World. 7 (1): 9-12.

7. Sadamoto, A. and Nobuyuki, Y. 1989. A Whistling Doneness Indicator for Microwavable Foods. Microwave World. 10 (4): 11-13.

8. Schiffman, R.F. 1987. Performance Testing of Products in Microwave Ovens. Microwave World. 8 (1): 12-15.

9. Southern, S. 1990. Prospects for Microwave Markets in Europe. Presentation at INSKO Conference. Finland.

TABLE 1

Microwave Oven Penetration of Households* 1980-1990 (percentage)

Country	1980-1983	1984	1985	1986	1987	1988	1989	1990
UK	6.0	11.0	17.4	25.0	33.0	40.9	46.1	50.7
France	1.2	1.8	3.0	6.0	11.5	18.9	26.4	34.1
W. Germany	1.3	2.0	3.3	6.7	12.2	20.5	28.5	35.2
Netherlands	0.2	0.5	1.1	2.3	4.9	9.5	15.9	22.6
Belgium	0.5	1.1	2.0	3.2	5.4	10.5	16.6	24.2
Italy	1.4	1.8	2.4	3.0	3.8	5.0	6.6	8.4
Spain	0.2	0.4	0.6	1.0	1.9	3.9	6.5	9.4

*   Note Average number of households 1986 (million):

UK	21.8	Netherlands	5.4
France	20.9	Belgium	3.7
W. Germany	26.4	Italy	18.7
		Spain	10.8

Source : Marketpower Estimates (Southern 1990)

TABLE 2

Effect of Shape on Mean Internal Temperature (°C) and percentage Weight Loss for Potato Samples of 2.6cm³ Volume

Cooking time (seconds)	Highest temperature (°C)		% Weight loss	
	Sphere	Cylinder	Sphere	Cylinder
5	47.07	31.05**	1.01	0.99
10	67.81	45.40**	2.74	2.88
20	71.98	51.46**	14.99	10.37**
25	70.99	52.16**	19.54	13.95**
30	66.80	51.70*	25.16	14.40**

** = significant difference p=<0.001

* = significant difference p=<0.005

(BURNETT, 1990)

TABLE 3

Effect of Shape on Mean Internal Temperature (°C) and percentage Weight Loss for Potato Samples of 12cm³ Volume

Cooking time (seconds)	Highest temperature (°C)		% Weight loss	
	Sphere	Cylinder	Sphere	Cylinder
5	42.20	26.42**	0.09	0.17
10	75.07	31.81**	0.28	0.23
15	85.74	43.96**	1.30	0.43**
20	84.50	56.41**	2.97	0.59**
30	88.39	66.68**	5.96	1.39**
40	91.53	76.19**	10.70	3.80**
50	90.40	75.55**	16.45	7.14**

** = significant difference $p = < 0.001$

(BURNETT, 1990)

FIGURE 1

PENETRATION OF MICROWAVES
2450 MHz

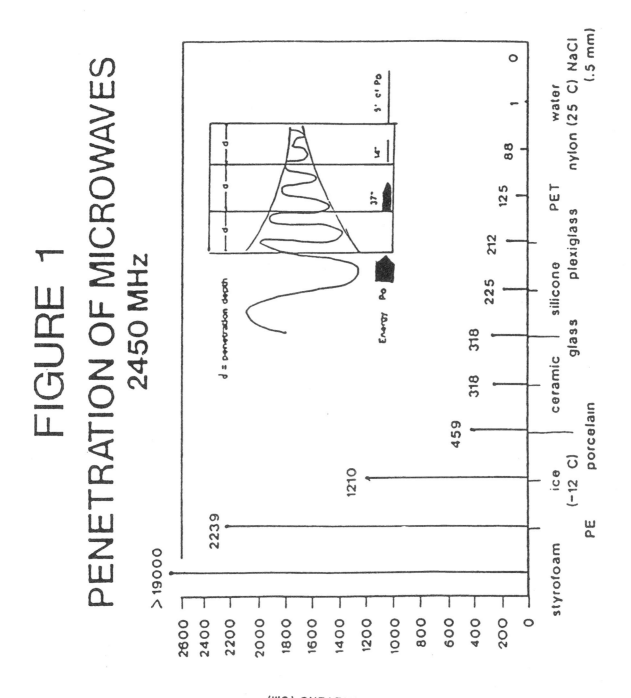

FIGURE 2

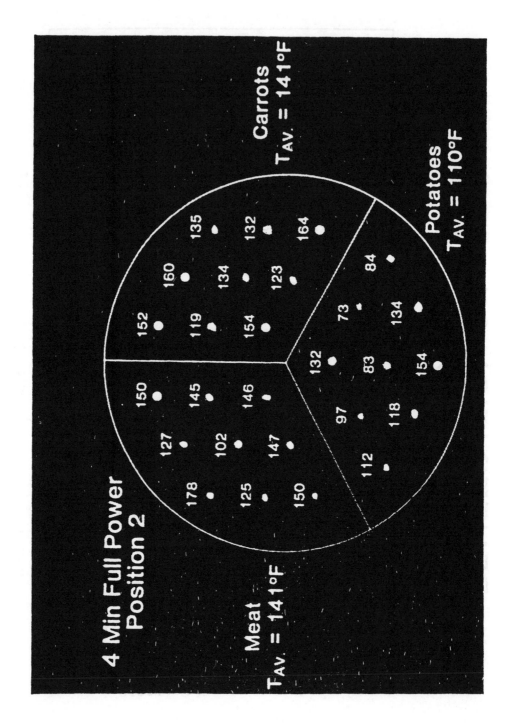

Typical temperature distribution immediately after heating.

(O'MEARA AND REILLY, 1986)

DEFINING AND DEVELOPING THE ENVIRONMENTAL PERFORMANCE
OF PLASTICS PACKAGING

John Salladay
Director, Environmental Performance
Plastics for Packaging Industry Group
Dow Plastics
P.O. Box 515
Granville, OH 43023

Biography

John Salladay is the Director of Environmental Performance and New Business Development for Dow's Plastics for Packaging Industry Group (PPIG) in the U.S. The Environmental Performance Group has responsibility for gaining an understanding of present and future environmental performance requirements for Plastics for Packaging and to establish the strategies and programs to meet those requirements with its customer partners along the packaging chain.

A twenty-two year veteran of Dow, Mr. Salladay has served in a variety of technical and commercial positions in product and applications development. He held management positions in Dow's Technical Service and Development organizations for STYROFOAM^R products and for Films and Coated Metals businesses. In 1987, he accepted the challenge to help Dow establish the methods and disciplines for market oriented product development and innovation in PPIG. He went on to lead Dow's "Solid Waste Assessment Team" in 1988 to provide Dow management with an understanding of emerging trends in solid waste management and their influences on package choice.

Mr. Salladay graduated with B.S. and M.S. degrees in chemical engineering from Ohio State University in 1968. He and his wife and four daughters reside in Granville, Ohio.

Abstract

With so much publicity on packaging and the environment, it has become increasingly difficult for packaging suppliers and users to know what steps to take to establish the positive "environmental performance" for their products. This paper identifies some of the emerging components of a definition for environmental performance in packaging and provides examples of recently developed packages that illustrate these principles. In addition the paper suggests the critical importance of communicating the functional as well as the environmental performance benefits of packaging to consumers.

Defining and Delivering
the Environmental Performance
of Packaging

John Salladay\
Director- Environmental Performance
Plastics for Packaging Industry Group
Dow Chemical U.S.A.
October 1990

A small group of people in Dow Plastics almost three years ago began a quest. The search was to understand current beliefs and practices in waste management and how they influenced plastic and plastic packaging. The study then grew to include the overriding trend of environmentalism.

It quickly became clear that within our search for understanding definitions were elusive: changes were taking place faster than we or others could absolutely define them. The calls for "environmental friendliness" and "sustainable development," to name but two, continued to be abstractions for the world, in spite of some very fine efforts to define them.

Language and law, however, can be powerful tools, so even as these evolving concepts were being refined, industry was called upon to fulfill them. And we continue to be asked to do so . . . . .

These spurs from government, the public, and various interest groups have been perceived by some to be threatening. Others take the opposite view, claiming that vast opportunities lie within the posturing, rhetoric, and realities about the protection of our planet. Almost no one suggests that there is no problem, and most have come to understand the validity of many environmental imperatives.

Our search began to yield dividends when we grasped the importance of the hierarchy of waste management, established by the United States' Environmental Protection Agency to deal with hazardous wastes some years ago. That very same hierarchy would now be applied to municipal solid waste. This small but significant point led us to identify early the push toward recycling. And our deepening understanding of source reduction, reuse, and recycling has helped guide the company as it develops its environmental policies.

As our small group has become larger within Dow we have seen the need to truly define and act upon a real requirement for the industrial community: **environmental performance.**

Dow's efforts have included some very special projects and programs: the development of a Materials Recycling Facility, an Applications

Development Laboratory, and a recycling program undertaken for several of the National Parks in the United States, to name a few.

We have acted in concert with the industry, too, in establishing the Council for Solid Waste Solutions, the National Polystyrene Recycling Company, the Polystyrene Packaging Council and much, much more.

But Dow and the plastics industry--and certainly plastic packaging itself--have much more to demonstrate in this area. It is very much evident that we must continue to define real environmental performance and then act appropriately upon our understanding.

Nevertheless, it is not the only material, nor are these the only industries that must finally address the delicate balances of growth and conservation in our world.

Aside from the chance to meet eminent colleagues and enjoy this lovely setting, are not these some of the important matters that draw you here today?

And yet it is not the *definition* of environmental performance that we most need, is it? Wouldn't the sheer weight of environmental evidence and concern demand the <u>demonstration</u> of that performance? Aren't we bound by a sense of profitability and responsibility to deliver the consumer and our world what they most need?

Now as I develop this talk you will see emerging a thesis. It is a thesis which draws upon the wonderful synergies among our industries. It applies to plastic. It applies to glass, to paper, to aluminum, and to steel. It applies to packaging. And most significantly, you have told us, it applies to you in the food business.

So even as you begin to appreciate the complexities of defining environmental performance, you will hear why, and in what context, that performance is so crucially important for us all to deliver.

Here is the thesis:
THE PRODUCERS AND USERS OF PACKAGING WILL ONLY GAIN RECOGNITION FOR PACKAGING'S ENVIRONMENTAL PERFORMANCE, WHEN THE PUBLIC REGAINS ITS APPRECIATION FOR THE PURPOSES OF PACKAGING ITSELF.

That's right. We must discuss packaging's overall performance, even as we demonstrate its environmental soundness.

Gone are the days when a new package can so conclusively demonstrate its value that the public leaps to buy whatever is packed in it. To illustrate this, let me tell you a brief story. It's one that's been told before, but it bears repeating here.

In the late 1800s in the United States travel by train was more popular than it is now. It was one of the best ways to travel cross country.

On one such ride a man named Samuel Crumbine made an amazing discovery. Trains were not air-conditioned then, unless you could call the hot, dusty breeze blowing through open windows conditioned air. People, frankly, needed to drink, especially in our western states where the temperature is unusually hot and where this story takes place.

The early version of the drinking fountain--a bucket full of water with a single cup--rested forward in the car, and passengers would make their bumpy way up the aisle to get a drink.

Crumbine, with some disinterest, watched an old and consumptive man drink from the cup, hacking and coughing between each sip. A few moments later a young child reached for the handle, only to have it ripped from her hands by Crumbine.

He had made the association among disease, contagion, and common containers--and so was born the disposable paper cup.

That's a story of *revolutionary* progress. It was a discovery of such magnitude--and with such obvious benefits to mankind--that one would think it difficult to forget.

There are other stories, too.

Imagine my grandfather's surprise and delight, for instance, when he no longer had to carry his beer bucket to the corner store! A glass bottle probably seemed to him a minor miracle.

No more flies on the pickles. No more bugs in the crackers. Candy without ants--these were unmistakable advances in packaging. Although I am sure there were some who longed for a return to the good old days, many would have rejoiced over the prospect of fewer stomach pains, less dysentery, and the longer lasting freshness of the foods they bought.

In these examples, and in countless others, the purposes of packaging were made abundantly clear to the consumer.

Startling advances in packaging almost always addressed HUMAN NEEDS. They addressed them in a way that made a lasting impression on the consumer. Where non-differentiated, bulk distribution of food was, there suddenly existed the *personal* package. Furthermore, the result was better health with little extra cost.

In the years between then and now--with all of their consequential events--the consumer has forgotten why foods are packed the way they are. In fact, in some cases, there may be fully two generations who have never understood exactly what packaging does for mankind.

But here is the truth: the demand for environmental performance in packaging is but another way of saying, "I have a personal need; help me fill it." All the heavy rhetoric from industry and environmentalists alike misses the central point--after years and years of fulfillment, the public perceives its basic human needs to be threatened. It believes that the industry developed to meet their needs has now, in fact, ignored them. And its ultimate faith in science leads them to believe that packs can and must be made to degrade, or they must be reused, or they must be recycled.

Complicating these factors is the long-held public belief that when you throw something away, it goes away--forever. Now that this myth has been broken, and now that the spectres of air and water pollution hang heavily in view, and now that it is recognized that packaging visibly contributes to the growing stream of waste, the public justifiably wants solutions.

Add to these points the charge of wastefulness. A central tenet of environmentalism has to do with the patterns of consumption that have developed with civilization's progress. It is widely put forward that packaging itself directly relates to those "false" patterns, try as it does to "sell" merchandise.

Without a shred of defensiveness, though, every material, every pack type, every application has its story to tell. And the part we all miss is VALUE.

Yes, of course packaging means to sell product. Its importance as a billboard within various markets has grown in recent times. But that is not the central purpose of packaging.

Even as environmental performance is raised to the level of function, form, and quality in packaging design, we are all missing the point--and I repeat myself--the aggravated assault against packaging for its environmental performance partially stems from a vast ignorance of its real purpose.

But let me fulfill my promise to you. I want to offer a definition of environmental performance. I then want to suggest the remarkable variety of attributes packaging already possesses. Next, I will align existing performance with anticipated performance.

I will then discuss an emerging means of measuring how pack materials and types perform, as well as a way to deliver to the world the central message of packaging's real value.

Environmental Performance

Packaging design, including as it does the challenges of functional performance and cost effectiveness, and no matter what the package or market, must now include environmental considerations. The contents of the package will continue to dictate certain functional requirements and in some cases will suggest the best waste management alternatives.

But what will this package look like?

It may be made of a single high performance material or of multiple materials that are readily recyclable. It will be manufactured using the least amount of energy possible, and it must include the least amount of material in meeting or exceeding its functional requirements. It could well be reused indefinitely. If it fails, or comes to the end of it useful life, it will be recycled and, in an ideal system, those recycled materials will form the new package or other packages. It will achieve all of these elements at a fair price.

When it is discarded, its energy value when burned will recapture most of the resources it used throughout its life. If the package at last makes its way to the landfill, the package will occupy little space, will leach no toxic substances, and will remain stable and benign as long as it stays there.

Presently, no such package exists. Many approximate these criteria. Some do particularly well on several. But no package completely integrates all these environmental elements with the highest sense of functionality and design.

Should we insist that it do so, ironically, we may fail to provide the public what it really most requires. But we will surely miss that "significant enhancement" to our quality of life that a careful integration of the most appropriate functional and environmental attributes will deliver.

The Attributes of Packaging

When we consider all the extraordinary functions that packaging performs, it seems almost criminal to ask for still more.

Nevertheless at the heart of these public requests are a depth and sincerity of purpose well beyond the ordinary.

What makes it so?

The standard of living in developed countries has risen dramatically in recent years. So much so, in fact, that most of our new products stress enhancing the quality of life, well beyond the goal of meeting basic human needs. Even as developing countries continue to wrestle with issues of health, sanitation, preservation, safety, and hygiene-- all related to shelter, warmth, and the food chain--so we refine the concepts of ready meals, debate the virtues of MAP/CAP, and work for ways to maximize machine speed.

Without attempt at social comment, the disparity between these ways of life is enormous--and enormously instructive.

Environmentalism galvanizes the return of a concern in developed countries for our basic human needs and virtues. All of a sudden the water we drink, the air we breathe, the crops we eat, and what we consume are all at issue.

What underdeveloped countries daily face--and with an intensity we can only approximate--we are having to once again confront. The correlation with packaging is not direct, but associative and symbolic. We come to represent for the public and its media the very antithesis of what we are.

Packaging means better health. It means containment and protection. It reduces food spoilage and preserves perishables in untold ways. Packaging means less waste.

Packaging also means safety, and through tamper prevention to a degree controls society's more aberrant behavior. Furthermore, packaging not only allows us to offer sensible selling messages, but it provides a common platform to communicate with our customers about what is inside the package.

I take the time to mention these more obvious points simply because our ultimate customer never knew them, or perhaps has forgotten them, or does not understand their importance. We can, in actual fact, claim he will appreciate their significance, if properly told, because his concern for his own well-being is high.

How Packaging Performs Today

It would be foolish to assert that packaging has met every requirement of every industry it serves. The changing tastes and demands of their respective consumers alone would change the requirements for each pack over time. What I can assert is that packaging fulfills its avowed purposes: to protect, contain, preserve, prevent, and inform; and to promote health, sanitation, safety, and hygiene.

How it does all these things in relation to the advances called for by changing needs or technology is easier to understand than how it performs environmentally. We have determined over the course of time, for instance, how to test, using what criteria, each one of these requirements. For example, consider the matter of insulation. If the object of the package is to retain heat or cold, we can measure their loss. We can measure the amount of time that passes in relation to that loss. We can compare these losses in relation to other test samples. And we can make determinations about what works and what does not.

If the object of the package is to cushion from shock, we can easily determine how well it does so by shaking it vigorously, by handling it roughly, or by dropping it from a high tower. We furthermore can compare those results with other competing systems.

My point here is not so much to develop that packages can be tested successfully by application and in comparison with others. Rather, it is to illustrate two things. First, that the requirements of a package are in large measure determined by its contents. Second, that science has developed accepted, if not always uniform, means of measuring performance in given circumstances. Acceptable results can be as simply stated as "It didn't break," to the more complex "The ambient temperature of the contents remained below -5C for 46 minutes, when tested in a sealed container exposed to a constant +25C."

Seen in this light, packaging performs. The matter of temperature extremes, resealability, clarity, moisture resistance, rigidity or flexibility--this is the stuff of science. As are oxygen barrier, taste and odor, durability, and retortability. We understand what causes certain bacteria to form under what conditions. Moreover, we know how to prevent their formation using additives, temperature control, or pack type. Again, THIS IS SCIENCE.

Today's packaging was created without environmental performance as a key criterion of design. We have a whole generation of packages, in fact, busily addressing the highest needs of the consumer, that were created well before the advent of environmentalism.

The environmental attributes of the great majority of today's packages are ancillary to the major purpose for their design. We more or less look beyond the form of the package to the material it employs. We measure its environmental interaction according to how it accommodates the hierarchy of waste management we've developed. And the various materials are assumed to perform well or poorly according to shifting criteria. THIS IS NOT SCIENCE.

To be fair, we have tested, "*using scientific procedures,*" how materials and packs reduce waste or do not, how they can be reused and over what periods of time, if they can be recycled and under what conditions, and if they require incineration or burial--and when.

But these products and packages, for the most part, have not been designed with a refined definition of environmental performance in mind. At best, they only hit part of the mark, because that mark has been invisible or at the very least fast moving.

So today's packages, fulfilling as they do their functional criteria, addressing as they do the very basic human needs of health and safety, often fall short of the mark when considering the environment. Complicating matters still more are the forgotten attributes of that packaging and the ill-defined term of **environmental performance.** Providing yet another twist are the as yet undetermined scientific elements within the measurement of that performance.

And we have ignored for the moment the entire public motivation behind environmentalism: right at the heart of it is--you've heard it before--**human need**.

Finally, though, we are where we are today because we have failed as an industry to **communicate** the value of what we do to a world that needs to hear it.

Let's turn now to the more hopeful signs of progress--and to some of the things we all can do to advance the science of environmental testing and communication. I'll do so by more specifically discussing the advent and future of ecobalance studies, but the message should be clear to all: we cannot separate the environmental performance of a material from its overall performance within any given application.

Life Cycle Assessments

The study of ecobalances, increasingly called life cycle assessments, has been with us for many years. Often using readily available data, the analyses mean to shed light on how products or process perform right from the extraction of raw materials to their ultimate disposal. Their focus is often on how energy is used within a product's life and how it may be recaptured through extended reuse or recycling. Credits are added or subtracted for what resources are used, how they are extracted, the additional energy that is required within manufacturing or fabrication, and how the product is used and for how long.

Although this simplistic discussion in no way does justice to the difficulty of accurately measuring how a package performs environmentally, there are few experts who would disagree that such a tool is valuable. To date, these various studies have employed scientific methodology, but the specific methods and scope of each study have differed substantially.

Nevertheless, it is within these studies that we find great hope. Several major studies are underway right now, and their early results look far more conclusive than previous studies. Moreover, there is an effort underway to set clear standards for measuring total environmental and life cycle performance.

A group of 54 experts in life cycle assessments--representing city, state, and federal governments, industry, academia, public interest groups, and research laboratories in the U.S., Canada, Japan, and several European countries--recently met to define current methodology and develop a technical framework for ecobalances.

"One of the most significant findings of the workshop," their news release states, "was the recognition that the complexity of environmental issues requires a more sophisticated approach than just quantifying materials and energy . . . as has been done in previous life cycle assessments."

The outcome of the workshop, conducted by the Society of Environmental Toxicology and Chemistry, is a three component model for performing LCAs:

1) an inventory of materials and energy used, and environmental releases (air, water, and solid waste) from all stages in the life of a product or process from raw material acquisition to to ultimate disposal;

2) an analysis of potential environmental effects related to the use of resources (energy and materials) and those releases; and

3) an analysis of the changes needed to bring about environmental improvements for the product or process under study.

LCAs until now have focused only on the first step within this model.

Once developed and used, this model can have far-reaching consequences for material choice and pack type, but even more important is how information is eventually shared with the public. These are points around which industry can rally. Standardizing tests, and then communicating their results in an understandable and unified way, can do much to enhance our credibility.

Even today there is news unfolding within the world of source reduction, reuse, and recycling that is difficult to share but entirely relevant for people to know. These messages must be sent in non-defensive, non-confrontational ways that address the human needs of the consumer.

Demonstrating Environmental Performance

The existing database of information demonstrating how packaging performs functionally in the world is ready to use. It must be shared with the public in simple but powerful ways.

Much of that information, particularly in food packaging, has to do with meeting basic human needs. We have seen in the environmental movement that the public reacts when it perceives the basic building blocks of life to be threatened. Its beliefs must eventually translate to real action, and we see continuing evidence that the public is indeed acting as it makes environmental decisions in the supermarket. Is it not reasonable to assume that the public will react favorably as we address environmental requirements within the context of their ultimate safety and health?

The emergence of environmentalism--and its subsequent focus on how packaging substantially contributes to the waste stream, how glass manufacturing consumes great energy, how paper in reality does not degrade rapidly in today's landfills, how aluminum still has 45% to go to achieve full recycling, how plastics use up valuable resources, and how steel makers pollute--simply points to yet another challenge for our respective industries.

But it provides us *an opportunity to speak about the value of packaging* and what it contributes to our world. As life cycle assessments unfold the real story about our respective materials, processes, and packages, these messages, wisely used, can lend credibility to our story.

Finally, though, we are where we are today because we have failed as an industry to COMMUNICATE consistently the value of what we do for a world that needs to hear it.

(That world, by the way, is uniquely unprepared to hear the recitation of only fact or science. Its needs are expressed, and understood, in the context of emotion. The science we perform--the facts we possess--are immaterial, if they cannot be communicated with feeling.)

We are all in the business of satisfying human needs. The attributes of our materials and packages are real and help satisfy those needs. The growing understanding of how our materials and packs perform in the environment can increasingly contribute to the fulfillment of safety and health, even as they already do.

So I ask for you to join with Dow in standardizing tests for environmental performance and communicating wisely the benefits of materials and packages that have long been forgotten or gone unnoticed. With good science we can truly add environmental performance to the long list of how packaging meets human needs. Only in these ways can we then continue to enhance the quality of life for our generation and the next generations to come.

We at Dow are committed to work with you in this endeavor.

Thank you.

T - #0695 - 101024 - C0 - 280/208/21 - PB - 9780877628668 - Gloss Lamination